McQueen 4.8.50
132531

Field Guide to the Peat Mosses of Boreal North America

Field Guide to the Peat Mosses of Boreal North America

Cyrus B. McQueen

University Press of New England
Hanover and London

The University Press of New England

is a consortium of universities in New England dedicated to publishing scholarly and trade works by authors from member campuses and elsewhere. The New England imprint signifies uniform standards for publication excellence maintained without exception by the consortium members. A joint imprint of University Press of New England and a sponsoring member acknowledges the publishing mission of that university and its support for the dissemination of scholarship throughout the world. Cited by the American Council of Learned Societies as a model to be followed, University Press of New England publishes books under its own imprint and the imprints of

Brandeis University
Brown University
Clark University
University of Connecticut
Dartmouth College
University of New Hampshire
University of Rhode Island
Tufts University
University of Vermont
Wesleyan University

Printed in the United States of America

♾

Library of Congress Cataloging-in-Publication Data
McQueen, Cyrus B.
Field guide to the peat mosses of boreal North America / Cyrus McQueen.
p. cm.
ISBN 0-87451-522-X
1. Peat mosses—United States—Identification. 2. Peat mosses—Canada—Identification. I. Title.
QK539.S75M38 1990
588'.1'097—dc20 89-40615
CIP

5 4 3 2 1

This book is dedicated
to my parents, Arlene and William McQueen,
who nurtured my love
of nature.

Contents

Preface

My interest in peat mosses was sparked by a naturalist in Algonquin Provincial Park, Ontario, Canada, in the summer of 1957 during my family's first major camping trip. I was only six years old, but I remember the moment as though it happened yesterday. We had gone for a nature walk with the naturalist and a group of other campers to a bog near the campground. The surroundings were unlike anything I had ever experienced near our home in suburban southern Ohio. The naturalist introduced us to pitcher plants, pink orchids, cotton grass, cranberries, and the carpet of "red" *Sphagnum.* He had us jump up and down on the soft turf. Incredibly, the ground quaked, and the trees bobbed up and down, making me laugh. But I also remember being slightly scared by this unexpected event. I came to regard "bogs" as mysterious gardens of nature; in fact, they still hold a certain mystery for me.

I wrote this field guide for two reasons. First, in the past ten years there has been a renewed interest in *Sphagnum* by professional botanists and amateur naturalists. I have been repeatedly asked by botanists and amateurs to identify their collections and to teach them about field identification of *Sphagnum.* At the present time, there are no field guides to the common species of *Sphagnum* like there are for other plant groups, in spite of the fact that peat mosses cover approximately 500 million acres of North America! The books on *Sphagnum* that are currently in print require the use of a compound microscope to key out the plants. This fact obviously limits the usefulness of these texts. I realized that for nonspecialists the information should be put into a format that can be readily used and understood. After all, the common *Sphagnum* species are no more difficult to recognize than any other species of plants.

My second reason for writing this book was in response to a professional concern. Although there has been a renewed interest in studying peatlands, especially by ecologists, many studies focus only on the vascular plants found there. Peat mosses, however, are excellent indicator species, and they can yield significant information on the water and nutrient conditions of a peatland. Also, the peat mosses are, in terms of biomass, the dominant vegetation in peatlands! Even if ecological studies only indicated what section of the *Sphagnum* genus was present in a peatland it would be an improvement over simply lumping all the species together into one *Sphagnum* category. Even small peatlands can hold up to 20 or more *Sphagnum* species!

Connected to my second reason is my concern over the fact that few taxonomy courses include the bryophytes (mosses and liverworts), in spite of the fact that they outnumber the ferns in numbers of species (approximately twenty thousand species worldwide—about twice the number of fern species). Many regions have at least a few hundred species of mosses alone! I include the bryophytes with the vascular plants in my taxonomy courses, and I have found that the students do not have any more difficulty identifying them than they do vascular plants.

This field guide is not meant to be an exhaustive

taxonomic treatment of the genus. Indeed, some species are difficult if not impossible to positively identify in the field without examining microscopic features. I have selected 26 species that I feel are common or distinctive enough to be recognized in the field. However, I mention nearly all the species found in boreal North America and have provided a key to all of them at the end of the book. The species not receiving a detailed treatment in this book can be identified if the specimens are exemplary (*anisophyllous* forms, see section on Morphology and Terminology). The 26 species that are closely examined cover the great majority of the acreage blanketed by *Sphagnum* and account for the species most frequently collected by nonspecialists. My hope is that this field guide will serve as only an introduction to this fascinating genus and its habitat for amateurs, students, and professional biologists. Happy bog trotting!

Acknowledgments

Many friends and fellow bog trotters knowingly and unknowingly helped me with this book. I want to thank especially Ian Worley, who was my graduate advisor at the University of Vermont. He made it possible for me to devote my professional career to the study of *Sphagnum* with his unfailing encouragement, support, and friendship. It was he who originally gave me the idea to write this field guide. Dick Andrus, who reviewed an early draft of this book, helped fine tune my field identification skills. His zealous pursuit of *Sphagnum* is contagious and has always been a source of inspiration for me. I would also like to thank Aaron J. Sharp who reviewed an early draft of the manuscript. Howard Crum informally edited some of the first identification keys that became a part of this book and was very helpful in answering my taxonomic questions. Some people were helpful because they accompanied me in the field and trusted me to teach them more about *Sphagnum*. I probably learned more about field identification from these experiences than from any other: David Barrington, Jerry Jenkins, Deborah Overrath, Cathy Paris, Liz Thompson, and Peter Zika. I would also like to thank the students of my plant taxonomy, natural history of Vermont, and peatland ecology classes at Johnson State College for testing many of my field keys. The fact that nonbiology majors could learn to recognize different species of *Sphagnum* in the field convinced me that a nontechnical book would be useful to those interested in natural history.

Pictures are worth a thousand words. The color plates in this field guide were selected for their instructional value more than their aesthetic sense, showing more than what I could put into words. I tried to use photographs that would be helpful in field identification. Most were taken in natural settings, but a few were posed to enhance their most distinctive characteristics. Bruce E. Perkins of Johnson, Vermont, offered invaluable technical advice on photography and helped me improve the quality of the photographs.

This book would not have been possible without aid from my family. Although they were at times unwilling bog trotters, my wife, Cheryl, and daughter, Shannon, have accompanied me in the field many times over the years and continue to do so. Cheryl's devotion to me is a constant source of comfort and inspiration. Shannon has been my "faithful assistant" since she was in first grade and has probably visited more peatlands and knows more of the natural history of peatlands than many professional botanists. Her keen eyesight and observational skills have helped me locate many elusive species.

C.B.McQ.

Johnson, Vermont
December 1989

Field Guide to the Peat Mosses of Boreal North America

Regions in North America to which this manual refers.

Introduction

The purpose of this guide is to make field identification of the different species of peat mosses easier. A combination of simple keys, color photographs, line drawings, and nontechnical descriptions are used to accomplish this goal. Approximately half of the species known to exist in the boreal regions of the United States and Canada are included in this field guide. These species are the most common ones encountered in the wetlands and peatlands of continental North America. Since the majority of the species included here have a circumpolar distribution, the guide will also be useful outside the United States and Canada.

The equipment needed for successful use of this field guide are inexpensive and easy to use. A hand lens of 10× to 20× is desirable to examine the plants, tweezers will aid in dissection, and a small metric ruler will be needed for taking measurements. An inexpensive field microscope can be made by taping together a discarded microscope objective and ocular using electrical tape. A permanent marking pen might be desirable for staining the leaves to ease identification.

Morphology and Terminology

The peat mosses belong to the genus *Sphagnum*. This is the sole genus in the class Sphagnopsida. As with all bryophytes (mosses and liverworts in the division Bryophyta), these plants have a dominant gametophyte (haploid) phase and a shorter-lived sporophyte (diploid) phase.

Figure 1. Branch fascicles. Two spreading branches and two pendent (hanging) branches, with stem leaves outlined at base of fascicle (left); and six spreading and four pendent branches (right).

The peat mosses differ from other mosses in the way their branches are organized on the plant. Branches are arranged in clusters called *fascicles* along the erect stem. Each fascicle consists of two or more *spreading branches* and one or more *pendent branches* (figure 1). The number and kind of branches in each fascicle are important clues in species identification. In this guide, for each species I have given the typical number of spreading and pendent branches per fascicle in a numerical shorthand. For example, a plant with two spreading and one pendent branch per fascicle is written as "2 + 1." Parentheses around a branch arrangement in a species description, such as "(2 + 2)," indicate a possible variation of the most common arrangement.

Young branches are usually crowded together at the top of the plant to form a head or *capitulum* (figure 2). As the stem elongates, the branches be-

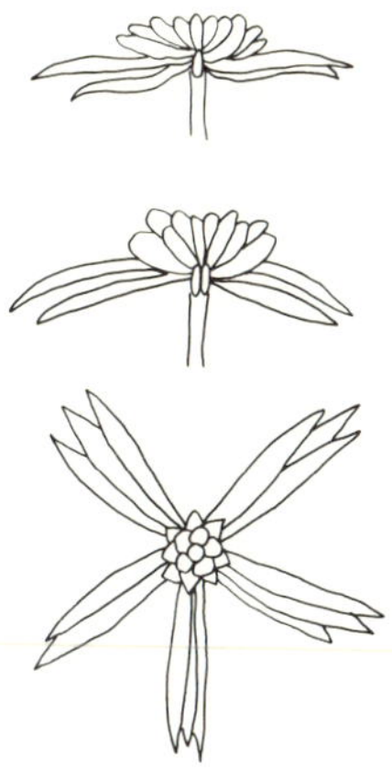

Figure 2. Stellate capitulum. Single branch between rays of capitulum as seen from side of species in section *Acutifolia* (top). Paired branches between rays of capitulum as seen from side of species in section *Cuspidata* (middle). Five-rayed (stellate) capitulum as viewed from above (bottom). (See plates 5 and 15 for examples of rounded capitula.)

come arranged in fascicles. The shape of the capitulum is a useful tool in separating some species. For example, the capitulum might be hemispherical, rounded, or flat-topped. Flat capitula often have five rays of branches radiating out from the center when viewed from above, a condition known as *stellate* or *five-rayed.* A compact *terminal bud,* from which new branches arise, might also be visible in the center of the capitulum (see *S. riparium* and *S. fimbriatum*).

There are two types of leaves on mosses. Leaves on the branches of the fascicles are called *branch leaves.* These may be arranged loosely on the branch or organized in five distinct rows on each branch, a condition called *five-ranked* (figure 3). My

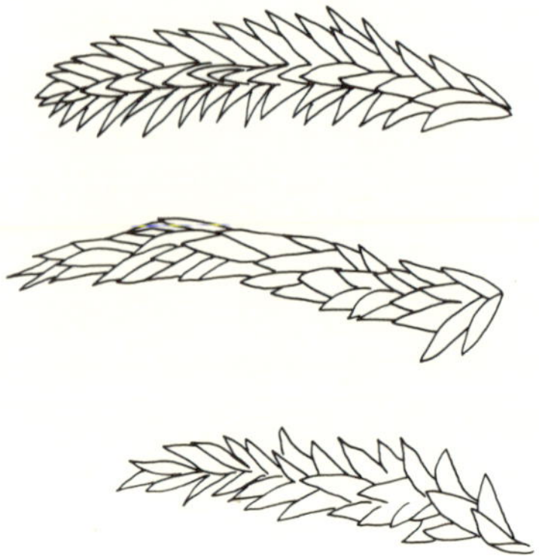

Figure 3. Branch leaf arrangement on spreading branches. Leaves arranged in five rows or five-ranked (top); imbricate leaves (middle); loosely imbricate leaves (bottom).

descriptions of branch leaf shape are taken from leaves off the lower half of the spreading branches. *Stem leaves* are found along the main stem (figure 1). These are more useful than branch leaves in species identification. It may be necessary to remove the capitulum and some branches in order to see the stem leaves. These leaves are often shaped differently from the branch leaves. A black or blue permanent marking pen can be used to stain the leaves so that they are more distinct. Examine the stem leaves from the upper living portion of the plant, since the leaves from the dead lower portion tend to be ragged and torn. The terms for describing leaf shape are defined in the glossary.

The level of difficulty in identifying species often depends on the degree of differentiation of stem leaves from branch leaves. By differentiation, I mean the degree which stem leaves differ in shape from the branch leaves. A single species may have an anisophyllous, hemiisophyllous, or isophyllous form. Anisophyllous forms have stem leaves that are maximally differentiated from the branch leaves. These plants are usually found in optimal habitats. They usually exhibit all the distinctive features of a particular species and are the easiest to identify. Hemiisophyllous forms have stem leaves that are only partially differentiated from the branch leaves. The existence of these forms indicates that the plants there are under some hydrologic stress. Not only is the microscopic morphology of hemiisophyllous forms different from anisophyllous forms, but so are the branch leaf arrangement, the number of branches per fascicle, the capitulum size and shape, and sometimes the color. Plants in which the stem leaves are not differentiated from the branch leaves are called isophyllous forms. These plants are usually found in habitats where there are frequent changes in water levels. The capitula of such forms might be very small, the branch leaf arrangement and number of fascicles can be different from the anisophyllous forms, and the branch leaves are often five-ranked, recurved, or squarrose, even in species that do not normally exhibit these features. The isophyllous forms of very distinct species can look identical. Although the section of the genus might be recognizable, it is often very difficult to key out isophyllous forms to species, even with the aid of a microscope! Welcome to the wonderful world of *Sphagnum* taxonomy.

Figure 4. Gametophyte and sporophyte (excerted on pseudopodium) of *S. centrale*.

The Life Cycle of *Sphagnum*

The *Sphagnum* plant is referred to as a *gametophyte* (figure 4) because it bears microscopic sexual organs that produce sperm and eggs, known collectively as *gametes*. The branches that produce the sperm, in structures known as *antheridia*, are found near the capitulum of the plant. The antheridia are found in short, club-shaped spreading branches near the base of the capitulum. These branches are often pigmented differently from the rest of the plant, with perhaps red or brown. The antheridia can only be observed with a microscope. It is impossible to find an egg-producing structure, called an *archegonium*, since they are very small. These are located at the base of some of the spreading branches near the capitulum. They are usually surrounded by branch leaves that are nearly twice

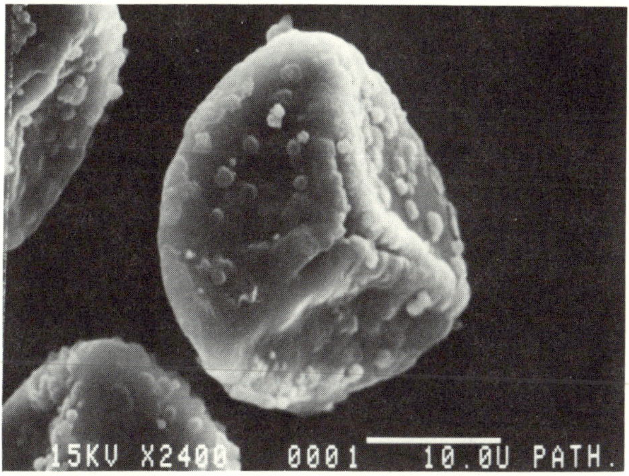

Figure 5. Photomicrograph of *S. capillifolium* var. *tenellum* spore taken with a scanning electron microscope. Bar = 10 μm (micrometers).

the size of the rest of the leaves. Some species of *Sphagnum* have the male and female structures on the same plant, *monoicous* species, whereas other species have separate male and female plants, *dioicous* species.

A fertilized egg develops into a spore-producing structure in a further phase of the life cycle. This *sporophyte* (figure 4) includes a *capsule* that is hemispherical in shape and dark brown to almost black in color. The capsules are only a few millimeters in diameter and are raised above the plant on small green stalks called *pseudopodiums*. As the capsules dry out, they explode with an audible "pop," shooting spores inside the capsule away from the plant. Each spore (figure 5) can germinate, producing a new green plant (gametophyte). Some species do not reproduce sexually as often as others, so sporophyte-bearing plants are not common. Capsules, if produced, are observable in the summer.

Habitat does seem to affect the timing of sporophyte production by different species. In New England, I have observed species that occur in more minerotrophic habitats, such as poor and medium fens (see discussion of these wetlands in the chapter on ecology), produce sporophytes in the early part of summer, usually in June. Some of these species were *S. magellanicum*, *S. fallax*, *S. teres*, *S. majus*, and *S. fimbriatum*. Most of these species grow near the water level and in exposed, treeless habitats. Species characteristic of ombrotrophic bogs and the drier portions of poor fens, such as *S. fuscum*, *S. capillifolium*, and *S. angustifolium*, seem to produce sporophytes in the middle part of summer, usually in July. Woodland species, such as *S. russowii*, *S. girgensohnii*, *S. subtile*, *S. centrale*, and *S. squarrosum*, produce sporophytes in the late portion of summer, usually in August. Of course, there is considerable variation caused by precipitation and local weather. I have mentioned this pattern to bryologists in Great Britain, Norway, Sweden, Finland, and Denmark, and they have apparently observed patterns much the same. This information is useful if you want to find sporophytes, for the pseudopodium dries out very soon after the spores are discharged and the sporophytes fall off the plants. On some species, such as *S. fimbriatum*, which may grow six or more inches in a summer, old sporophytes might still be found on the lower portions of the plants.

The most common mode of propagation by *Sphagnum* is by vegetative reproduction. Gametes are not involved in this process. In vegetative reproduction, the top portion of the plant might produce a branch that eventually develops into a stem with a capitulum and fascicles of branches. After several seasons of growth, the older portions of the branch decay, resulting in two separate plants. This mode of reproduction is undoubtedly important for many species, since sporophytes are not often produced by them. Of course, some species, such as *S. fimbriatum*, produce abundant numbers of sporophytes every season, with some plants having two or more sporophytes. But other species, such as *S. capillifolium* in wetter habitats, do not often produce sporophytes and use vegetative reproduction.

Vegetative reproduction can also occur by a fragment of a plant growing into a mature plant. This mode of reproduction is common in many mosses. Asians have obviously known of this phenomenon for a long time, for they start their "moss gardens" by grinding up plants into small fragments and spreading them on damp soil. I have used the same principle for cultivating *Sphagnum* in the greenhouse. I chop a single plant into many fragments and spread them on some moist peat. After about a month, many small plants are visible, each arising from one of the fragments.

Vegetative reproduction is probably important for long-distance dispersal of *Sphagnum*, although no one has examined this issue in any detail. A few observations in the field, however, suggest that it does occur. It is common to find individual plants of *Sphagnum* pulled out of the peat and left drying on the surface of a *Sphagnum* mat or hummock. It is likely that small animals, such as birds or even mice, pull out the plants. Peat mosses are not known to be a food source for any animal, but perhaps the plants are being used in nest-building, or perhaps they are being pulled out by accident. It is doubtful that humans are responsible, because in many of these places there is no evidence of human disturbance. Besides, there are not many people currently studying *Sphagnum*. Small fragments of these plants might be carried off by animals or blown away by the wind.

Not all species survive long after being pulled out of the peat. Species such as *S. capillifolium* do not grow well if they have been dried for more than one week. However, I have been able to get dried specimens of *S. fallax* and *S. angustifolium* to grow after being dried for more than two months after I collected them.

The question of how *Sphagnum* gets established in new habitats has not been studied. This problem is intriguing, since it has been documented that in some quarries in the Midwest abandoned 20 to 30 years ago significant *Sphagnum* populations are now found. It is my hope that this field guide will help amateurs and professional ecologists and biologists study some of these questions by providing a means to identify some of these widespread species.

Ecology

A basic understanding of the ecology of peat mosses will be useful in identifying the different species of *Sphagnum*.

The geographical region covered by this field guide corresponds to the North American boreal and subarctic taiga, a broad transcontinental crescent from Newfoundland to Alaska with minor subalpine extensions moving southward into the Appalachian Mountains in the eastern United States. Pleistocene glaciation has been largely responsible for modifying the topography to favor the development of *Sphagnum*-dominated wetlands. Since these wetlands are very common in the Arctic regions, and are found on the tops of mountains in the tropics, one can conclude that their distribution is controlled by other factors as well, such as temperature. The *Sphagnum*-dominated wetlands composed of boreal species found in Michigan, Illinois, Indiana, and Ohio indicate that they are relicts of former climatic conditions.

Vast areas of the northern hemisphere are covered with peat mosses. These plants play a significant role in controlling drainage in northern regions, which might otherwise be subject to considerable erosion. Peat mosses are found in wetland habitats throughout the world, on every continent except Antarctica. The greatest number of species occur in North America and northern Europe; however, very little is known of *Sphagnum* outside these regions. (This ignorance results in part from the fact that the professional taxonomists studying

Sphagnum are located in North America and Europe.) The great diversity of peat mosses in these regions is no doubt due to the poor drainage in the glaciated portions of the northern hemisphere.

Peat mosses become established in poorly drained wetlands because such habitats are often too poor in nutrients to support other vegetation. Vascular plants, such as the flowering plants and conifers, simply outcompete *Sphagnum* in wetland habitats where nutrients are plentiful. Peat mosses in general have very slow growth rates, although there are a few species, such as *S. fimbriatum* and *S. recurvum*, that can grow a foot per year under optimum conditions. *Sphagnum* is one of the few species that can grow in poorly drained wetlands low in nutrients.

Habitat Zonation of *Sphagnum*

Peat mosses are often referred to as "indicator species." Such a designation by ecologists suggests that the discovery of a given species in an area infers additional information about that habitat because the species exhibits specific habitat preferences. This is not true of all peat mosses, because some species are "generalists" and are found in a wide range of habitats. A few of the common species covered in this field guide, however, are excellent indicator species. In fact, one reason I wrote this guide was to help increase our knowledge of *Sphagnum*-dominated wetlands by providing a simple way for amateurs and professional biologists to identify common species of *Sphagnum*. By knowing what species occur in different *Sphagnum*-dominated wetlands, it is possible to make predictions concerning the hydrology and nutrient levels of those habitats, because different species occurring at various levels above the water table exhibit different preferences for various nutrients. Conversely, it is also possible to predict what species will be found in a particular area if the type of wetland is properly recognized. For this reason, I later discuss the different types of *Sphagnum*-dominated wetlands found in boreal North America.

Anyone who takes an interest in peat mosses will begin to notice that the species are arranged along various gradients. The wet-to-dry gradient, the nutrient gradient (most easily examined by determining the pH), and the gradient based on the degree of exposure to sunlight are relatively easy to examine.

Ecologists have long recognized the relationship between different species and their height above the water level. Species also differ in their tolerance for desiccation, their nutrient requirements, and their ability to carry on photosynthesis under various conditions. I will devote the majority of the discussion on these factors to the relationship between different species and height above water level because this is the simplest gradient to study and because there is considerably more information available on it from ecologists. It is also of greater utility to the users of this field guide. Of the other gradients, I will mention nutrient requirements in a broad sense and mention the rest only briefly since very little information is available on them and they are not as valuable as aids for species identification.

Although peat mosses are found primarily in wetland habitats, very few of the species are truly submerged aquatic plants. The majority of them grow near or even as much as several feet above water level in a wetland. The diverse kinds of wetland habitats in which peat mosses are found no doubt accounts for the large number of species—estimated at 150 to over 300—in the genus. The disagreement over numbers of species stems from the fact that, like most aquatic plants, individual species exhibit considerable morphological variation, some species being more variable than others. Some, such as *S. russowii*, that grow in many different types of wetlands are sometimes difficult to recognize in the field since they superficially resemble other species. In such cases, microscopic examination of the plants are necessary to identify them with any degree of certainty. Throughout the species descriptions in this field guide, I have made note of such morphologically variable species and given hints for recognizing them in the field without the aid of a microscope.

Zonation of species along the wet-to-dry gradient. The morphology of peat mosses enables them to retain large amounts of water. The large, empty hyaline cells of the leaves and the branches along the stem, especially the pendent branches, can hold large amounts of water. Water can also move by capillary action on the outside of the plant due to the wicklike nature of the pendent branches. These features enable peat mosses to withstand desiccation and actually to grow above the water table. The various species of *Sphagnum* are adapted to growth at different heights above water level, and this adaptation is reflected in their macroscopic and microscopic morphology. Many of the common species are distinctive enough to be recognized in the field, hence this field guide.

Even though there is considerable overlap among species in their habitat preferences, a definite wet-to-dry gradient may be observed for most species. This gradient is also known as the hummock-hollow sequence, in reference to the relatively drier mounds or hummocks of peat mosses and the wet hollows closer to or at water level.

Several morphological trends related to different heights above water level are observable in *Sphagnum.* Species growing submerged or near the water level often have very weak stems, the plants are very limp, the fascicles of the branches are widely spaced along the stem, and the capitulum is flat and not well developed. Species typically found in drier habitats have more rigid stems, the fascicles of the branches along the stem are usually more closely spaced, and the capitulum is usually more developed and rounded in shape. Species growing above water level are also more likely to contain other colors besides green or may be brightly colored. However, these trends must be accepted as only very general tendencies. It is also possible that some species that are normally above the water table during the summer are submerged during the wetter months of the spring and fall. These changing conditions may have an overall effect on the general appearance of the plant and may affect the spacing of the branch fascicles and the rigidity of the stem.

There are several wet-to-dry species sequences, for not all species are found in all *Sphagnum*-dominated habitats. Also, some species overlap, and some wet-to-dry sequences are more apparent than others. The sequences I will discuss here are the most common and easily recognized. Other wet-to-dry sequences of peat mosses may exist; certainly not all have been studied.

The most common wet-to-dry species sequence is the one found in the *Sphagnum*-dominated wetlands that most people think of as a true bog. (Since I will be discussing the different kinds of wetlands later in this chapter, I will now give only brief descriptions of the different types.) Bogs are *Sphagnum*-dominated wetlands whose water supply comes strictly from atmospheric precipitation. These environments are the most acidic of the *Sphagnum*-dominated wetlands. The sequence of common species found in bogs from wet-to-dry is *S. cuspidatum—S. majus—S. fallax—S. angustifolium—S. magellanicum—S. capillifolium—S. fuscum*. This sequence can be observed over even a short distasnce, from the edge of a bog pond to the top of a nearby hummock, from the bottom of a hollow to the top of an adjacent hummock, or might be discoverable over a long distance through a large, raised bog. Thus, in some cases the sequence is easily witnessesd, while in others a considerable distance must be covered to observe the whole sequence.

The species of this sequence illustrate very clearly the effect of water on the morphology of these plants. *Sphagnum cuspidatum* is a very weak-stemmed plant, with broadly spaced branch fascicles and a relatively undeveloped capitulum. Unlike the other brightly colored species of this sequence, this primarily aquatic species is usually green. The emergent *S. majus*, which can grow submerged, has a stronger stem, more closely packed fascicles, a prominent capitulum, and ranges in color from pale green, dark green, brown, or even blackish. *Sphagnum fallax* is a much more compact plant, with a more rigid stem, closely packed fascicles, a prominent capitulum, and a greenish or pale color with some tints of orange and yellow. The next species in the sequence, *S. angustifolium,* is similar to the previous species but is smaller, often more compact, and has a flush of red or pink in the stem. *Sphagnum magellanicum* is one of the largest species in the genus. It has a very stiff stem, many large fascicles, a prominent capitulum, and is deep red in color. *Sphagnum capillifolium* is a smaller plant that often packs the fascicles so tightly together that it is difficult to separate them. This species has a very large capitulum and is bright red in color. Note, though, that one variety of this species that grows closer to the water level, *S. capillifolium* var. *tenellum,* has a smaller, flat capitulum. The species occupying the driest portion of the bog, usually on the tops of large hummocks, is *S. fuscum.* This species is small, with a flat capitulum, and dark brown in color. This sequence is certainly one of the more colorful ones.

The wet-to-dry species sequence in the *Sphagnum*-dominated wetland known as a fen is similar to that of the bog. A fen is a *Sphagnum*-dominated wetland whose water supply has some nutrient input from minerals in the soil, that is, the water supply is not strictly atmospheric. Most people call such wetlands bogs, but they are different in a number of respects. In general, they are wetter, have more nutrients, and are not as acidic as true bogs. The sequence of common species from wet-to-dry is *S. cuspidatum—S. majus—S. papillosum—S. fallax—S. recurvum—S. subsecundum—S. angusti-*

folium—S. magellanicum—S. fimbriatum—S. teres—S. capillifolium. The major difference between this sequence and that of a true bog is the inclusion of a few species that are characteristic of wetter habitats, higher pH, and higher nutrient levels. *Sphagnum fuscum* does occur in fens, in fact it is common in some drier, more acidic portions of fens, but overall it is not as common as in a true bog.

The wet-to-dry sequence of species in swamps is not as clearcut as in bogs or fens. A swamp is a forested wetland that is either flooded during part of the year or has ground water flowing through it. The close proximity of the water table to the forest floor no doubt accounts for the lack of a distinctive gradient. Swamps can be composed of hardwoods or conifers. The latter might consist of either spruce or cedar. Cedar swamps are richer in nutrients and less acidic than spruce swamps and are frequently referred to as treed fens. The following sequence includes species from different types of swamps, so not all of them would be present in a single swamp: *S. centrale—S. russowii* = *S. warnstorfii* = *S. girgensohnii—S. squarrosum—S. subtile—S. wulfianum*. The equal (=) sign between species indicates that these species overlap so much that it is difficult to place them in any absolute order. It should be noted, however, that *S. warnstorfii* is a species of cedar swamps and some rich fens and is usually not found in association with *S. russowii* and *S. girgensohnii*. The latter two species, on the other hand, are often found growing intermixed with one another.

The wet-to-dry gradients that have been presented should not be thought of as absolutes but rather as tendencies of species. Local and seasonal variations in the water level have an effect on the position a species will occupy along this gradient. It is common for many species to be submerged for at least brief periods during the spring and fall when there is a greater amount of precipitation. These sequences only typify what is usually observed during the summer when most people, professional biologists performing research or amateurs out for a hike, visit *Sphagnum*-dominated wetlands. It is apparent that not all species included in this manual are mentioned with regard to these sequences. I again emphasize that some species are more general in their habitat preferences and may be found throughout any one sequence.

These wet-to-dry sequences are useful in several ways. First, they may be useful in the identification of a particular species. For example, if a plant is found near the top of a hummock, there is very little chance that it is one of the more aquatic species such as *S. cuspidatum* or *S. majus*. Likewise, a brown plant growing on top of a damp hummock several feet above the water level is almost certainly *S. fuscum*. Second, once a species is identified, it is possible to estimate how far away the water level is from the surface of the *Sphagnum* mat. The distribution of species along the wet-to-dry gradient is determined by the availability of water throughout the year, thus, one can get a sense of the relative position of the water level on a seasonal basis. If, for example, there is very little change in species over a long distance, such as a large mat of *S. magellanicum* with some *S. angustifolium* mixed in, one can conclude that the water level is relatively stable over that distance. If, on the other hand, the wet-to-dry sequence is repeated again and again over a short distance, one can conclude that there is considerable variation in the water table.

Zonation along a nutrient gradient. In addition to its ability to soak up and retain large quantities of water, *Sphagnum* can also acidify a habitat by exchanging hydrogen ions (H^+) for other ions such as potassium, sodium, calcium, and magnesium (K^+, Na^+, Ca^{++}, and Mg^{++}, respectively). The degree of acidity is quantified by using a pH (potential hydrogen) scale. This scale is logarithmic, ranging from 1 to 14, with each whole number representing a change in the power of 10. Acidity is a measure of the H^+ concentration, with a pH of 1 being the most acidic, 7 being neutral (the concentration of H^+ is equal to the concentration of OH^- [hydroxide ions]), and 14 being strongly basic or alkaline (highest concentration of hydroxide ions, for example, soaps and bleach).

The majority of ecosystems are close to neutral in pH, however *Sphagnum*-dominated wetlands are invariably acidic, sometimes measuring below a pH of 3. The ability of *Sphagnum* to acidify the environment has a number of advantages. First, the plants use this exchange of H^+ for other ions as a means of securing them, since they are necessary for growth (they are often referred to as nutrients) and are in low concentrations in this habitat. Second, by creating an acidic habitat, peat mosses produce an ideal environment for themselves where there is little competition from other plants, especially from vascular plants. Most vascular plants cannot tolerate the poor nutrient levels or acidic environments of *Sphagnum*-dominated wetlands;

therefore, this ion exchange reduces competition for the peat mosses.

The molecule responsible for cation exchange (the exchange of positively charged ions), galacturonic acid, is located in the cell walls of peat mosses. Galacturonic acid consists of a galactose molecule in which the primary hydroxyl group is oxidized to a carboxyl group. The carboxyl hydrogen is weakly held, thus it is readily available for exchange. In the exchange process, hydrogen ions (H^+) are released, and metal cations (for example K^+, Na^+, Ca^{++}, and Mg^{++}) are adsorbed in the carboxyl site on the galacturonic acid molecule. The buildup of these H^+ in the environment accounts for the acidifying properties of *Sphagnum*.

In general, peat mosses invariably lower the pH of the environment that surrounds them. There is, however, a strong correlation between galacturonic acid content of different species and optimum height above water level. Species toward the dry end of the wet-to-dry sequence, such as those growing near the tops of hummocks (*S. fuscum* for example), generally have more galacturonic acid than species that grow near water level (for example, *S. cuspidatum*). The amount of galacturonic acid within the plants of each species confers upon each a specific range of exchange capability and, therefore, assigns each species to a particular niche.

Using this information, it is possible qualitatively to describe a pH gradient based on the wet-to-dry sequence. In general, the environment at the wet end of the sequence will be less acidic (closer to 7) than that at the dry end of the sequence (closer to 1). Note, though, that this gradient is only qualitative. The pH of the habitat for individual species has been found to vary as much as 1 pH over the course of a year due to fluctuations in precipitation. In general, the lowest pH values in *Sphagnum*-dominated wetlands are reached during the summer. The pH increases in the spring and fall when there is a greater amount of precipitation. As can be observed from table 1, which lists species and pH measurements, including range (lowest and highest values) and mean (average, using at least 10 measurements), there is a broad overlap among species. These pH measurements were taken from *Sphagnum*-dominated peatlands in Maine, New Hampshire, and Vermont over the course of several summers. They are thus representative of only a small portion of the range of peat mosses and for a single season, that is, during the summer. Nevertheless, if one finds a species such as *S. warnstorfii*, it can be inferred that the environment there is very weakly acidic. Generalizations about local variations within a single wetland are also possible, so that it may be stated that one particular part of a wetland is more acidic where, for example, *S. fuscum* is common than another part where *S. magellanicum* is common. This information is also useful when comparing the acidity of different wetlands. If the majority of the species in one wetland are typical of more acidic habitats and the majority of species in another wetland are typical of less acidic habitats, then one can conclude that the first wetland was more acidic than the second.

The pH gradient combined with the wet-to-dry gradient is helpful in estimating the relative concentrations of ions in the water. It would be expected that more acidic, drier portions of *Sphag-*

Table 1. Ranges and means of pH for species from New England

Species	pH	
	Range	Mean
S. angustifolium	3.6–6.1	4.2
S. capillifolium	3.5–5.2	3.9
S. centrale	3.5–7.8	6.0
S. cuspidatum	3.5–5.9	4.3
S. fallax	3.5–6.1	4.1
S. fimbriatum	3.4–7.0	5.2
S. flexuosum	3.5–6.3	4.2
S. fuscum	3.5–5.1	3.8
S. girgensohnii	3.9–5.8	4.5
S. magellanicum	3.5–6.2	4.5
S. majus	3.6–5.1	4.3
S. papillosum	3.6–6.3	4.6
S. pulchrum	3.8–6.2	4.4
S. riparium	3.6–6.0	4.8
S. russowii	3.8–6.1	4.5
S. squarrosum	4.6–7.3	6.1
S. subsecundum	4.7–6.2	5.6
S. subtile	4.3–5.9	4.9
S. teres	4.6–7.3	5.9
S. warnstorfii	4.1–7.2	6.3

num-dominated wetlands are poorer in nutrients than the less acidic, wetter portions. This expectation is also useful for the hummock-hollow gradient, where the top of the hummock is drier and usually more acidic than the wet hollow and the hollows are usually richer in ions than the hummocks. No absolute limits can be given, however, since there is considerable local variation in ions in the water and in seasonal precipitation. For example, coastal regions will probably have a greater concentration of Na^{+}, K^{+}, and Mg^{++} than inland habitats because of the greater concentrations of these ions in the oceans. Ion concentrations can also vary radically from day to day, depending on the amount of precipitation received in a wetland. Precipitation can dilute the concentrations, or it may actually increase the concentration locally.

Zonation based on degree of exposure. Individual species of *Sphagnum* vary in their tolerance for sunlight. Species such as *S. centrale*, *S. subtile*, *S. squarrosum*, *S. girgensohnii*, *S. russowii*, and *S. wulfianum* are found only in shaded habitats. Other species, such as *S. fuscum*, *S. capillifolium*, *S. magellanicum*, *S. papillosum*, *S. angustifolium*, *S. subsecundum*, *S. majus*, *S. cuspidatum*, and *S. fallax*, are rarely found in a forest. Some species grow in shaded habitats under shrubs or at the edge of a forest bordering a *Sphagnum*-dominated wetland. Such species include *S. quinquefarium*, *S. palustre*, *S. teres*, and *S. fimbriatum*. This information can be used in identifying species, for the list of possible species is reduced if, for instance, a plant is collected in a forested wetland. However, as you become more familiar with the various species, you will no doubt find exceptions even to the exposure gradient.

The degree of exposure does have an effect on the color of some *Sphagnum* species. Many of the species that are exposed to direct sunlight are brightly colored. *Sphagnum capillifolium* and *S. magellanicum* are usually deep red, but may be green if growing under shaded conditions. *Sphagnum fuscum* is a beautiful dark brown, but in some shaded habitats, it may have a greenish tint. Likewise, *S. subsecundum* is brown to orange, but with a dirtier look to it. *Sphagnum angustifolium* is a light green with a pink to light red stem. *Sphag-*

num fallax may be green, under shaded conditions, to pale yellow or brown when growing in the sun.

Very few of the woodland species are brightly colored, but there are some exceptions. *Sphagnum warnstorfii*, for instance, has a dark, reddish purple color to it, and *S. russowii* and *S. subtile* are green speckled with red.

Some species, regardless of their degree of exposure to the sunlight, are green. Many of the near aquatic species such as *S. cuspidatum* and *S. majus* are primarily green in color. *Sphagnum fimbriatum*, which is often exposed to direct sunlight, is always green. Of the woodland species, *S. centrale*, *S. girgensohnii*, *S. squarrosum*, and *S. wulfianum* are nearly always green.

Of course, green specimens of every species can be found. This sometimes makes species determination difficult in the field without a microscope. *Sphagnum magellanicum* is usually red, but in shaded habitats it may be green and superficially resemble *S. centrale*. *Sphagnum russowii* and *S. girgensohnii* often grow intermixed but are usually easily separated by color, since *S. russowii* is speckled red and *S. girgensohnii* is always green. *Sphagnum russowii*, however, can be green without any tint of red, making it much more difficult to distinguish from *S. girgensohnii*. They are, however, easily separated on the basis of their stem leaves.

Peatlands

I have tried so far to avoid using the terms bog, fen, marsh, and swamp until I have described them in more detail. For that reason, I have used the somewhat tortuous term, "*Sphagnum*-dominated wetland," to refer to the prominent habitats in which the greatest diversity of *Sphagnum* is found. I normally refer to these different habitats by the term "peatland," for I feel that this term distinguishes these wetlands from the many other kinds of wetlands and that it describes a feature about *Sphagnum* that makes it unique among all other plants—that it forms a soil referred to as peat.

Peat is a type of soil that develops under waterlogged, anaerobic conditions. It has an organic content of at least 20 percent and an ash content of less than 50 percent. In reality, peat is very easy to recognize by its dark brown to black color, soft texture, and the presence of many undecayed organic materials, especially plant fragments. Peat can be formed by other plants, for example, by grasses and sedges, but I only discuss *Sphagnum* peat here.

Sphagnum peat accumulates as a direct result of the living plant. The growth of *Sphagnum* is primarily apical (at the top of the plant). As the upright stems elongate, the older portions below gradually die. Because the environment is very acidic, caused by living peat mosses exchanging hydrogen ions for other metal ions, the growth and reproduction of the bacteria and fungi that normally act as decomposers in ecosystems are limited or even prevented. Water also plays a major role by limiting the amount of oxygen available to decomposers. The combination of low pH and water prevents decay of the older portions of the *Sphagnum* plants. The growth rate of many *Sphagnum* species exceeds the rate of decay, thus peat accumulates. Peat deposits in the glaciated portions of North America that have been free of ice for ten thousand years vary in depth from a few feet to more than 50-feet deep. Obviously, some habitats are more conducive to peat formation than others.

Peatlands are classified in many different ways. In fact, professional ecologists often disagree over what names should be used and how many different kinds of peatlands should be recognized. Many of the classifications utilize plant associations within the peatland, topography of the peatland, and water chemistry. The primary problem in classification is that each peatland has some unique features determined by local geology, topography, and hydrology; however, some general classes of peatlands are easily recognized. I use a classification system that is of practical use to the amateur and professional alike, and one that will be useful throughout the boreal north.

Peat mosses grow in a variety of habitats often referred to as bogs. Actually, though, there are many different kinds of wetlands that are commonly called bogs, and it is useful to differentiate them. I thus recognize the following types of peatlands: the bog, fen, marsh, and swamp. There are several different categories of peatlands within each of these major types, not all of them dominated by *Sphagnum*. Depending on the available nutrients and water supply, different peatlands support very different groups of *Sphagnum* species as well as other types of plants.

Bogs. Bogs are wetlands that are very poor in nutrients because most of the minerals are obtained from precipitation rather than from ground water. They are also referred to as ombrotrophic peatlands, meaning simply that all nutrients come strictly from precipitation. They are usually characterized by a pH of 4 or less. They support an unusual group of plants such as pitcher plants, different orchids, sundews, cotton grasses, cranberries, and blueberries. *Sphagnum* is the dominant vegetation of a bog. The margins of bogs are usually very wet. Species that thrive in more mineral-rich habitats can be found here, such as *S. recurvum*, *S. magellanicum*, *S. fimbriatum*, *S. flexuosum*, *S. majus*, and *S. cuspidatum*. In the drier central region, where there are very few trees, hummock-forming species are common, such as *S. fuscum* and *S. capillifolium* (var. *capillifolium* and var. *tenellum*). In the wet hollows, *S. fallax*, *S. majus*, and *S. angustifolium* are common. In the wetter areas of the woodlands surrounding bogs *S. centrale*, *S. girgensohnii*, *S. subtile*, and *S. russowii* may be present. A moat, referred to as a *lagg*, surrounds the margins of many bogs where they abut the drier upland. Decay is rapid in the lagg because nutrients from the adjacent upland wash down into it; it thus can support plants other than *Sphagnum*. The assemblage of plants is then different from the vegetation in the rest of the bog. *Sphagnum* is not very common in the lagg, and species that do occur there are usually characteristic of more mineral-rich habitats. The lagg is usually characterized by vegetation reminiscent of a marsh, such as cattails, sedges, grasses, bulrushes, lilies, and pondweeds. It is ironic that the wettest portion of many bogs is its edge (at least, during the summer)! The lagg may be a few feet to many yards in width.

At least three different types of bogs are recognizable. *Raised bogs* are characterized by a domed central portion that is higher than the margins. These bogs form where yearly precipitation and humidity are high, such as along the coasts of Maine, Washington, eastern and western Canada, and Alaska. They can also be found further inland, however. Raised bogs are the most acidic type of peatland. *Blanket bogs* form under even more humid conditions. These peatlands follow the topography of the land and literally creep over the surface. There are very few blanket bogs in North America; they are more common in the British Isles and in northern Europe. Small blanket bogs are common on the tops of mountains that are frequently shrouded in clouds. These peatlands are usually slightly less acidic than raised bogs. *Flat bogs* are common in North America. These peatlands develop in depressions throughout most of the glaciated regions of North America. They are characterized by a flat surface. The peat in these bogs literally fill up the depression. Since the surface may be close to the water level, these peatlands are wetter than the other types of bogs. They are also less acidic.

Fens. Fens are very similar to bogs, but they receive some nutrients from the surrounding ground water and are wetter. Hummock-forming species such as *S. fuscum* and *S. capillifolium* var. *capillifolium* are not as common as in bogs, being replaced by *S. capillifolium* var. *tenellum* and *S. magellanicum*, which may form hummocks. Species found only at the margins of bogs might be distributed in pools and wet depressions throughout a poor fen. Many of the same plants that are found in bogs are also found in fens. Like bogs, fens can have laggs around their margins that support vegetation characteristic of more mineral-rich habitats.

Fens are often referred to as *minerotrophic*

peatlands because of the greater influx of nutrients from ground water. The term fen may be somewhat misleading, because the category includes some diverse habitats. For example, the word fen is used to describe some alkaline wetlands that are dominated by sedges and grasses and very few peat mosses.

Fens are probably the most common type of *Sphagnum*-dominated wetland in North America. They are characterized by a wide range in pH, from just above 4 to 7. Three broad classes of fens are recognized: poor, medium, and rich fens.

Poor fens have a flora very similar to that of a bog. They are most similar to a flat bog, with the exception that poor fens have an influx of nutrients from ground water. In general, they are wetter than bogs and have a pH of about 4 to 5. The species along the wet end of the wet-to-dry gradient are more common in poor fens. Thus, although species such as *S. fuscum* and *S. capillifolium*, so typical of bogs, are present, *S. cuspidatum*, *S. majus*, *S. angustifolium*, *S. fallax*, *S. flexuosum*, *S. papillosum*, and *S. magellanicum* are much more common. The hummock-hollow topography is also not as pronounced as in a bog.

Medium fens lack the species so typical of the drier portions of bogs and poor fens (*S. fuscum*, *S. capillifolium*, and *S. magellanicum*). Species that are found close to the water level and that prefer high nutrient levels are common, such as *S. teres*, *S. squarrosum*, *S. warnstorfii*, *S. centrale*, *S. subsecundum*, and *S. fimbriatum*. The nutrient levels are higher than in a poor fen, there is little or no hummock-hollow topography, and the pH is higher (5 to 6.5).

Rich fens receive a large influx of minerals from the surrounding ground water and are dominated by sedges rather than by *Sphagnum*. Species common to these wetlands are *S. centrale*, *S. subsecundum*, *S. teres*, *S. squarrosum*, *S. contortum*, and *S. fimbriatum*. Other plants familiar to bogs and poor fens, such as pitcher plants, are noticeably absent from rich fens. As with medium fens, hummock-hollow topography is absent. There is usually a large amount of standing water, and, correspondingly, the nutrient levels and pH (6 to 8) are higher.

Swamps. Swamps are forested wetlands in which there is standing water throughout most of the year. These wetlands support very few species of *Sphagnum*. An exception is the mineral-rich cedar swamp or, as it is often called, the cedar bog. Cedar swamps support many species of poor and rich fens in addition to *S. warnstorfii*. Of course, shade-tolerant species such as *S. centrale*, *S. russowii*, *S. girgensohnii*, *S. subtile*, *S. squarrosum*, and *S. wulfianum* are frequently found there.

There are many other types of habitats that *Sphagnum* may grow in. *Carrs* are alder- and willow-dominated wetlands common in flood plains. Small populations of *S. fimbriatum*, *S. teres*, and *S. squarrosum* may be found in these wetlands. Such habitats are usually not referred to as peatlands because decay is rapid enough so that peat does not accumulate. *Springs* or *seeps* in wet meadows often support a few *Sphagnum* species, such as *S. quinquefarium*, *S. subsecundum*, and *S. squarrosum*. As with carrs, peat does not accumulate. *Sphagnum* also grows along the shores of ponds or streams, in ditches, on wet banks, on wet cliff faces, and along trails or in disturbed wet areas at high elevations. These habitats rarely accumulate any peat.

Succession

Succession refers to the change in numbers and different kinds of species, leading eventually to a climax (stable) community. It involves the gradual replacement of one community with another. The change from one community to another may not be readily observed, since it may occur over several hundred to several thousand years.

Anyone who has seen a wide range of wetlands such as a pond, marsh, rich fen, medium fen, poor fen, and bog can easily visualize that this sequence is the path that leads to eventual bog formation. The sequence is not always followed, however, because local physiography and climate can alter the stages of succession or even reverse them. Actually, there is probably a broad overlap of communities as succession proceeds, so that at any one time the wetland may not fit into any of the artificial categories ecologists have created. A better description of bog succession may be that there are as many types of bog successions as there are bogs.

For example, I recently extracted a peat core from a small kettlehole bog near Stowe, Vermont, that was 6.5 meters in depth. The top 5 meters were solid peat, with *Sphagnum* spores and plant remains found throughout it. However, the lower 1.5 meters of the peat contained several alternating bands of fine silt with layers of peat separating them. *Sphagnum* spores and the remains of *Sphagnum* plants were found in these peat layers also. Obviously, succession was not very straightforward in this kettlehole bog. The original bog was submerged, only to fill in again, and this cycle was repeated over a period of perhaps two thousand years. This was probably a local exception, since I have not observed anything like it in the other kettlehole bogs in this region.

The stages of succession that I discuss in this chapter are based on my own observations in the field and from the published accounts of other scientists. I only present one account of succession that is possible; it is certainly not the only one. *Indeed, the zonation of different species that is often interpreted as succession in a bog may simply indicate local differences in water and nutrient conditions.* My intent is simply to provide a background in bog succession in a small pond to cue in readers as to what to observe. They may then make their own observations and come to their own conclusions. After all, bog succession in eastern North America is no doubt different from western North America, not only because different species and communites are involved but also because the environments are different.

Bog succession often begins around the margins of a lake or pond. As sediments are deposited around the margins through erosion, plants are able to invade the body of water. *Sphagnum* and other plants might invade the mat if conditions are suitable. In the case of *Sphagnum*, this means that the water is usually too poor in nutrients to support larger vascular plants. Accumulations of biomass, in the form of peat, permit the mat of vegetation to extend further out towards the center. After thousands of years, the encroaching vegetation meets in the center and encloses the pond or lake beneath it. Meanwhile, closer to shore, small shrubs and trees may start invading the drier portions of the mat close to shore. The continued accumulation of peat fills in the body of water and permits the trees to extend from shore to shore.

Does this scenario actually occur? I think in some cases it does occur; at least, I have made enough observations to convince myself that it does. The following example is based on succession in a small pond or lake. This succession sequence is known as the classic textbook Hydrarch Succession or Lake-fill Succession. I believe that this mode of succession can be divided into three stages, beginning with a preclimax community (bog) and ending with a climax bog (peatland) forest community.

The first stage is characterized by aquatic plant vegetation that grows along the shallow margins of a pond. These plants include *Potamogeton* ssp. (pond weed) (ssp. means some species from this genus, the species may vary depending on the region), *Nuphar* ssp. (yellow water lily), *Nymphea* ssp. (white water lily), *Typha* spp. (cattail), *Ranunculus* ssp. (buttercup), and *Scirpus* ssp. (bulrush). These are certainly not the only plants that occur in this first stage, but they represent the most obvious ones.

In the second stage, sedges, grasses, and rushes are more common in shallow water. These include *Carex aquatilis* and *C. diandra* for sedges, *Calamagroastis canadensis* (bluejoint grass) and *Glyceria canadensis* (rattlesnake grass) for grasses, and *Eriophorum* ssp. (cotton grass) for rushes. Some mosses might be found growing among the grasses and sedges, such as *Drepanocladus* ssp., along with a few peat mosses such as *Sphagnum subsecundum* and *S. teres*. As organic material accumulates along the margins, *Salix* ssp. (willow) become more common.

The third stage can be thought of as the actual *Sphagnum* invasion of the wetland. *Sphagnum recurvum* is one of the first species to invade the water. As the environment slowly becomes more acidic, *S. magellanicum* becomes more common. After *S. magellanicum* has had a chance to form small hummocks that rise above the water level, *Vaccinium oxycoccus* (small cranberry) becomes established on the smaller hummocks, and *Eriophorum angustifolium* (tall cotton grass), *Andromeda glaucophylla* (bog rosemary), and *Rubus chamaemorus* (baked-apple berry) begin to grow on the larger, drier hummocks.

At this point, the inner portion of the mat may be floating, with the outer edges filled in forming a solid deposit of peat. *Ledum groenlandicum* (Labrador tea) and *Eriophorum spissum* (hare's tail cotton grass) eventually invade the driest of the *Sphagnum* hummocks close to shore. Other *Sphagnum* species invade the mat, such as *S. capillifolium* on the exposed portions and *S. squarrosum* in the shaded portions. Eventually, small trees invade the mat close to shore. These include *Larix laricina* (tamarack), *Picea mariana* (black spruce), and some *Betula* ssp. (birch).

A bog actually develops after *Sphagnum fuscum* becomes established and forms hummocks on the mats of *S. magellanicum*. *Sphagnum capillifolium* and the rest of the peat mosses characteristic of a bog then appear. *Ledum groenlandicum* becomes the dominant shrub, and *Vaccinium* ssp. (cranberry) become more common.

The climax community of this ecosystem is a forest. Spruce, pine, and tamarack are the most dominant species of the bog forest. Birch may invade the drier margins of the forested peatland. As the peat mosses grow beyond their ability to draw up water by capillary action from the water table, they dry up and die, eventually being replaced by other mosses such as *Hylocomium splendens* (fern moss) and *Ptilium crista-castrensis* (plume moss). Mosses such as *Polytrichum* ssp. (hair cap mosses) and lichens such as *Cladonia* ssp. (including reindeer "moss" and British soldiers) begin to occupy the driest hummocks of *Sphagnum fuscum.* So, in the end, *Sphagnum* may be so successful that it literally "grows beyond its means"!

As I said in the beginning of this section, this is only one possible scenario for bog succession. Ecologists believe that in some cases climax communities are never reached, or at least not for an extremely long time. This may occur when the wetland is well drained, which may limit peat formation. Some ponds and lakes are also too deep for peat to effectively fill them in. In these wetlands, the action of ice in the winter and waves in the summer may erode the floating mat so that it can never close over the entire surface. Of course, one can not ignore the actions of mammals such as beavers, who may flood a peatland and cause portions of the peat to tear loose from the bottom of the depression and float to the surface. These bog islands can actually float around on the surface of the lake with the vegetation still intact, including shrubs, trees, and of course *Sphagnum.*

A climax forest may also fail to be achieved because the peatland is simply too wet. It is a common occurrence in poor fens for black spruce to invade the *Sphagnum* mat and to grow for 20 to 30 years, attaining a height of about 30 feet. After that time, the trees become so heavy that they sink down through the peat, become water-logged, and eventually die. Small branches at the base of the trees often grow into separate trees themselves after being overgrown at the base by *Sphagnum*. The dead trees in portions of these fens are usually the same height.

The reverse process can also take place, that is, a peatland may form by drowning, submerging, or simply overgrowing a dry upland. This is the exact opposite of Hydrarch Succession! The process is referred to as *paludification*. Paludification is undoubtedly the process responsible for some larger peatlands. It is very striking to come up on a coastal peatland where the *Sphagnum* has literally overtaken and blanketed a forest. Paludification could be initiated by rising water tables, a cooler, moist climate, or a combination of both. Hydrarch and paludification may have been responsible for the formation of some of the large, complex peatlands. It is beyond the scope of this book to discuss this process in any detail. I only hope that I have convinced the reader that the formation of peatlands is complex, no matter what theory is championed.

Using the Keys

I have included traditional and nontraditional keys in this manual in the hope of providing as much information as possible to aid in field identification of the various *Sphagnum* species. The first key, a random access key, provides a quick way of identifying species. It has its advantages and disadvantages; it is most appropriate for those not familiar with the more traditional keys. I have included a traditional key for identifying the section of the genus a plant is in. The user will then refer to the individual section keys to identify the species. This key is useful because it shows the evolutionary relationships among species in the genus.

I have also included a key to all the species at the end of the manual. This is an ambitious undertaking since the key includes nearly all the species found in North America, although I intentionally limited this manual to include only those species that can be identified with some certainty and that are the most common. Many of the species in this last key are highly variable and require microscopic observation. I have tried to use characteristics of plants growing in optimal habitats, where the morphology is the most distinctive. Of course, some species grow in a large range of habitats and may not always express their most distinctive characteristics. I recommend that the user try this key after having experienced some success in identifying species with the other keys. I have not included complete descriptions for all the species in North America, and the user may want to consult more technical manuals to confirm identifications of species not fully treated here.

Probably the most difficult sections of the genus in which to identify species are the sections *Sphagnum*, *Subsecunda*, and *Cuspidata*. The species of these sections are very common and grow in a broad range of habitats, so anyone who visits a *Sphagnum*-dominated wetland will probably find them there.

Random Access Key to Species

The random access key differs from the other keys in this field guide in that the user can select any characteristic(s) to begin the process, in any order, usually with the feature most easily observed by the user. The key uses 67 characteristics. Each species is represented by a number, 1 through 27. The list of numbers following each characteristic indicate which species *do not* exhibit this feature. To use the key, first write down the numbers 1 through 27 on a piece of paper. Next, select a characteristic and cross out the numbers that are listed after that feature. Repeat this procedure with another feature or features until one or only a few numbers are left on the list. If only one number is left, the description and photograph of the species should be consulted to confirm the identification. If a few numbers are left, either select one additional characteristic to eliminate more possibilities or look up each species in the field guide and select the species description that best describes the unknown plant.

For example, if a specimen has six or more branches per fascicle, there is only one possible species according to the random access key, *S. wulfianum*. The description and photograph can then be consulted to confirm the identification. For a second example, suppose a plant is "red." This characteristic eliminates all but numbers 2, 8, 11, 12, 15, 16, and 17. But the plant has branch leaves that are five-ranked, and this eliminates all but 11, 15, and 17. The plant has only 2 spreading branches per fascicle, eliminating 11. The plant is growing in an ombrotrophic bog, eliminating 15. Thus the possible identification is number 17, *S. capillifolium* var. *tenellum*.

It should be kept in mind that this key is not foolproof. Some species of *Sphagnum* exhibit a large range of variation, not all of which is understood by *Sphagnum* taxonomists. The list also includes less than half the species found in North America, so any given plant might not be included in the list. I have, however, included a discussion of the species not described in this manual in the descriptions of those species most similar to them in morphology. The list of characteristics is not exhaustive, but I believe it includes enough to enable you to identify all the species listed in this guide.

Species List for Random Access Key

1. *S. compactum*
2. *S. magellanicum*
3. *S. centrale*
4. *S. papillosum*
5. *S. wulfianum*
6. *S. squarrosum*
7. *S. teres*
8. *S. pylaesii*
9. *S. subsecundum*
10. *S. contortum*
11. *S. quinquefarium*
12. *S. fimbriatum*
13. *S. girgensohnii*
14. *S. russowii*
15. *S. warnstorfii*
16. *S. capillifolium* var. *capillifolium*
17. *S. capillifolium* var. *tenellum*
18. *S. fuscum*

19. *S. subtile*
20. *S. riparium*
21. *S. cuspidatum*
22. *S. majus*
23. *S. angustifolium*
24. *S. recurvum*
25. *S. flexusosum*
26. *S. pulchrum*
27. *S. fallax*

Characteristics *Not* Present in Species

1. Branch leaves
   a. cucullate (hooded); 5–27
   b. secund; 1–8, 11–16, 18–25, 27
   c. squarrose branch leaves; 2–5, 8–27
   d. recurved when dry; 1–19, 21–24, 26
   e. undulate; 1–24
   f. 5-ranked; 1–4, 6–10, 12–14, 16, 18, 20–21, 25
2. Number of spreading branches per fascicle
   a. 6 or more; 1–4, 6–27
   b. 3–5; 1–6, 8–10, 12–27
   c. 2; 5, 7–8, 11
   d. 0–1; 1–7, 9–27
3. Capitulum
   a. stellate with paired branches between rays; 1–19, 21
   b. stellate with single branch between rays; 1–10, 20–27
   c. rounded to hemispherical; 1–4, 6–10, 12–15, 17–18, 20–27
   d. prominent apical bud; 1–7, 9–11, 13–19, 21–27
4. Exposure
   a. exposed; 3, 5–6, 13–15, 19
   b. shaded; 1–2, 4, 7–10, 12, 16–18, 20–27

5. Color (all species may be green)
    a. red throughout; 1, 3–7, 9–10, 13–14, 18–27
    b. green tinged with red; 1–10, 12–13, 15–18, 20–27
    c. brown; 2–3, 5–6, 9–17, 19–21, 23–25
    d. red-brown; 2–7, 9–27
    e. brown-green; 1–4, 6–9, 11–14, 16–21, 23–27
    f. pink; 2–10, 12–15, 18–27
    g. pale green; 2–4, 7–11, 13–27
    h. yellow; 2–3, 5, 8–11, 13–20, 22–23, 25–27
    i. yellow-orange; 1–7, 9–27
    j. yellow-green; 1–8, 10, 12–19, 21–22, 26–27
    k. yellow-brown; 1–9, 11–22, 24
    l. orange; 1–8, 10–27
6. Ecology
    a. ombrotrophic bog; 1, 3–15, 19–20, 22, 24–25
    b. poor fen; 1, 3, 5–15, 19
    c. medium fen; 1–2, 4–5, 8, 10–11, 13–14, 16–19, 21–27
    d. rich fen; 1–2, 4–5, 8, 11, 13–27
    e. swamps or wet woodlands; 1–2, 4, 7–10, 12, 16–18, 21–23, 25–27
    f. carrs; 1–5, 8–11, 13–23, 25–27
    g. on bare sand and rock; 2–27
    h. wet cliffs and seeps; 2–7, 9–10, 12–27
7. Habitat
    a. hummock; 1, 4–11, 13, 20–22, 24, 26–27
    b. carpets and mats; 8, 18, 21–22
    c. wet hollows; 1–22
    d. submerged or at water level; 1–7, 9–20, 23–27

8. Stem leaf shape
   a. short triangular (equilateral triangle); 2–4, 6–8, 12–18, 20–22, 25
   b. triangular-lingulate (isosceles triangle); 1–4, 7–9, 12–14, 17–18, 23, 26–27
   c. lingulate; 1, 3–5, 8–12, 19–27
   d. ovate, widest above middle; 1–2, 4–27
   e. ovate, widest below middle; 1–3, 5–27
   f. ovate, widest at middle; 1–7, 9–27
   g. spatulate (fan-shaped); 1–11, 13–27
9. Stem leaf apex
   a. rounded; 5, 17, 19–21, 24, 26–27
   b. rounded with notch; 1–13, 15–27
   c. acute; 1–4, 6–10, 12–14, 17–18, 20, 22, 24–27
   d. fringed; 1–11, 14–27
   e. involute; 1–14, 16–25, 27
   f. split down middle; 1–19, 21–27
   g. truncate; 1–23, 25–27
   h. blunt; 1–22, 24, 26–27
   i. apiculate; 1–25
10. Stem leaf size
    a. <1 mm; 2–4, 6–8, 10–22, 24–27
11. Stem color (all species may be green; species not listed are green)
    a. red; 1, 3–13, 18, 20–27
    b. brown; 2, 5–7, 10–12, 14, 16–17, 19–27
    c. pale brown; 1–2, 4–7, 10–24, 26
    d. black; 1–3, 6–27
    e. red-brown; 1–4, 8–27
    f. yellow; 1–7, 9–27
    g. yellow-brown; 1–25, 27
    h. pale green; 1–9, 11–27
    i. greenish yellow; 1–10, 12–20, 23, 25–27
    j. pink; 1–22, 24–26

Plate 1. *Sphagnum compactum.*

Plate 2. *Sphagnum magellanicum.*

Plate 3. *Sphagnum centrale* (large-leaved, light green plant), *S. russowii* (red-speckled plants), *S. girgensohnii* (dark green plants with stellate capitulum), and *S. angustifolium* (yellow-green plants) growing in a forest at the edge of a peatland.

Plate 4. *Sphagnum papillosum.*

Plate 5. *Sphagnum wulfianum.*

Plate 6. *Sphagnum squarrosum.*

Plate 7. *Sphagnum teres*, squarrose form.

Plate 8. *Sphagnum teres*, nonsquarrose form.

Plate 9. *Sphagnum pylaesii.*

Plate 10. *Sphagnum subsecundum.*

Plate 11. *Sphagnum contortum.*

Plate 12. *Sphagnum quinquefarium.*

Plate 13. *Sphagnum fimbriatum.*

Plate 14. *Sphagnum fimbriatum* (small green plants) and *S. fallax* (large, pale green plants), both with young sporophytes.

Plate 15. *Sphagnum girgensohnii.*

Plate 16. *Sphagnum subtile* (red tinged plants with round capitula), *S. russowii* (speckled red plants with stellate capitula), and *S. girgensohnii* (green plants with stellate capitula) growing in forest at edge of peatland.

Plate 17. *S. warnstorfii.*

Plate 18. *Sphagnum capillifolium* var. *capillifolium* (densely branched plant with round capitulum) and *S. capillifolium* var. *tenellum* (plant with small, nearly flat capitulum and five-ranked branch leaves).

Plate 19. *Sphagnum fuscum.*

Plate 20. *Sphagnum riparium* (two large plants with prominant apical buds), surrounded by *S. fimbriatum.*

Plate 21. *Sphagnum cuspidatum.*

Plate 22. *Sphagnum majus* (brown plants), *S. cuspidatum* (yellow-green plants), and *S. magellanicum* (red plants).

Plate 23. *Sphagnum recurvum* (large, pale green plants with recurved leaves), *S. flexuosum* (plants with yellow-orange branch tips, pale green capitula, and slightly recurved branch leaves), and *S. angustifolium* (plants with yellow-orange branch tips and dark green capitula).

Plate 24. *Sphagnum recurvum* (right), *S. flexuosum* (center), and *S. angustifolium* (left).

Plate 25. *Sphagnum pulchrum.*

Plate 26. *Sphagnum fallax* with sporophytes.

Plate 27. *Sphagnum magellanicum* (large-leaved, red plants), *S. capillifolium* var. *tenellum* (small-leaved, red plants with stellate capitula), and *S. angustifolium* (pale yellow-green plants) growing on low hummock in open peatland.

Plate 28. *Sphagnum magellanicum* (large-leaved plants, green form), *S. fuscum* (small brown plants with stellate capitula), *S. capillifolium* var. *tenellum* (red plants), and *S. angustifolium* (pale yellow-green plants with stellate capitula) growing at base of large hummock in open peatland.

Plate 29. *Sphagnum austinii* (large-leaved plants with sporophytes) and *S. fuscum* (small-leaved plants) growing on large hummock in open peatland.

Plate 30. *Sphagnum bartlettianum* (red plants), *S. palustre* (large-leaved, pale green to yellow plants), *S. flexuosum* (pale green plants with yellow-orange branch tips), and *S. recurvum* (small-leaved, pale green plants with recurved leaves) growing in carpet in wet forest.

Descriptive Key

The genus *Sphagnum* is broken down into groups called sections. Species within a section share some characteristics. The section key should be consulted before searching for the species. After determining the section, then go on to the species keys under the individual sections.

Each key is composed of pairs of contradictory statements under the same number. Always start with pair number 1. Decide which statement most accurately describes the specimen and proceed to the next indented pair. For example, a specimen with noncucullate, squarrose branch leaves, four branches per fascicle, and lingulate stem leaves longer than 1 mm would key out to section *Squarrosa*. Go the species key of section *Squarrosa* to determine the species. Any unfamiliar terms should be looked up in the glossary.

The descriptions of the plants have been somewhat standardized. Each description contains information on overall aspect of the species, color, capitulum shape, branch arrangement, leaf shape, and habitat. I have also included a short discussion of characteristics that distinguish individual species from closely related ones. In cases of doubt, a careful examination of descriptions and illustrations along with a process of elimination will often result in identification.

For those stimulated by this book, a list of references is given in the "Further Reading" section at the end of this book. These technical treatments can be consulted for more critical identification. A microscope is necessary for successful use of these books, but they contain many excellent drawings and information on the habitats of different species.

Key to the Sections of *Sphagnum*

Please note: Not all species will key out to the proper section. For example, *S. teres*, which is in section *Squarrosa*, may have terete branches (cylindrical) rather than squarrose leaves. Not all sections in the genus are included in this key since it is designed for the common species described in this manual. A second section key is provided in the key to all species of North America at the end of the guide. Leaf figures' scale is 2.1 cm = 1 mm.

1. Branch leaves cucullate (figure 7)
    2. Stem leaves small and triangular, less than 1 mm long; branch leaves often tightly imbricate, somewhat squarrose—I. *Rigida* (*S. compactum*), p. 51
    2. Stem leaves large, greater than 1 mm long, lingulate or fan-shaped; branch leaves tightly or loosely imbricate, may sometimes be squarrose, especially in boreal zone—II. *Sphagnum*, p. 55
1. Branch leaf margins inrolled or flat, but not cucullate
    3. Six or more branches per fascicle (figure 1)—III. *Polyclada* (monotypic—*S. wulfianum*), p. 61
    3. Five or less branches per fascicle
        4. Branch leaves squarrose (figures 11–12)
            5. Stem leaves small and triangular, less than 1 mm long—I. *Rigida* (*S. compactum*), p. 51

5. Stem leaves large and lingulate, 1–1.5-mm long—IV. *Squarrosa*, p. 65

4. Branch leaves not squarrose
   6. Pendent and spreading branches very similar or plants unbranched, flexuose, and about the same length—V. *Subsecunda*, p. 71
   6. Pendent branches usually longer and more slender than spreading branches; the distinction between pendent and spreading branches is pronounced
      7. Stem leaves nearly the same size as the branch leaves or larger; plants various shades of green, brown, red; on plants with stellate capitula, branches of fascicles single or absent between the rays of branches (figure 2); plants of drier, mineral-poor, acid habitats; not normally submerged, usually growing above water level—VI. *Acutifolia*, p. 77
      7. Stem leaves much smaller than branch leaves; plants various shades of green, yellow, brown, never red, but sometimes branches and stem pinkish; on plants with stellate capitula, paired branches of fascicles are present between the rays of branches (figure 2); plants of wet, mineral-rich depressions; submerged or near the water level—VII. *Cuspidata*, p. 97

Section I. *Rigida*

The species in this section may superficially resemble species in sections *Sphagnum* and *Squarrosa*; however, the cucullate-appearing branch leaves of species in section *Rigida* differ from those in section *Sphagnum* in that the margins of the leaves are not joined at the apex to form a true hood. The stem leaves are also typically less than 0.8 mm long, much smaller than those of either section *Sphagnum* or *Squarrosa*.

S. compactum DC

(figure 6 and plate 1)

Short, stout plants, usually growing in compact cushions. Plants are pale green to pale yellow in color, rarely pink. Capitulum not well developed; individual shoots are often so crowded that the top, short, stout branches point upward, making individual capitula difficult to recognize. Stem is dark brown to almost black in lower portion of plant, light brown in upper portion. Fascicles of branches are close together, 2 + 2 (2 + 3); spreading branches short and swollen (1 cm or less in length); pendent branches longer than spreading branches and tightly appressed to stem. Branch leaves large (1.5–3.0 mm long), very concave, often appearing cucullate, sometimes slightly squarrose. Stem leaves very small (0.3–0.7 mm long), triangular-lingulate with rounded or broad lacerate apex.

Figure 6. *Sphagnum compactum* stem leaf (left) and branch leaf (right).

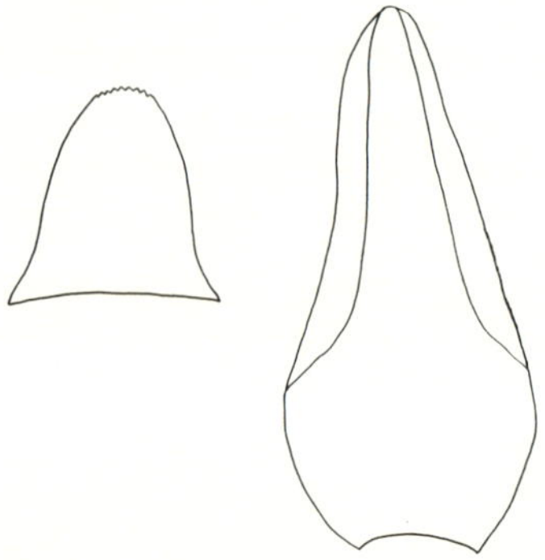

This species is often found growing on poorly drained sand, in wet drainage ditches, or on wet cliff faces. The compact, whitish green cushions of plants resemble "pin cushion moss," *Leucobryum glaucum* (Hedw.) Ångstr., but the presence of fascicles of branches, typical of nearly all peat mosses, differentiates it. A similar species, *S. strictum* Sull., may be distinguished in the field by its pale green to light brown stem, laxer habit, and strongly squarrose branch leaves.

Section II. *Sphagnum*

The distinctive macroscopic characteristic of species in this section is the cucullate branch leaf. The margins of the branch leaves are joined at the apex to form a hood. Most of the branch leaves are ovate and very concave. Species in section *Rigida* have similar branch leaves, but the margins are not joined at the apex. Some of the largest species of peat mosses are in this section. There are at least eight species in North America, but most of them are distinguishable only by microscopic characteristics, and so field identification is difficult. There are three common, easily distinguished species. Except for *S. magellanicum*, all identifications should be checked microscopically.

Key to Species

1. Stem leaves long rectangular, sides parallel; plants pink, red, purple-red, green—*S. magellanicum* (figure 7)
1. Stem leaves not long rectangular, sides not parallel; plants never red
   2. Stem leaves fan-shaped, widest above middle; plants green, gold-brown, purple-brown; forms dense carpets in wet forests—*S. centrale* (figure 8)
   2. Stem leaves ovate, widest at or below middle; plants most often gold-brown to purple-brown, rarely green; forms carpets in open areas, sometimes on rocks of mountains, but rarely found in forests—*S. papillosum* (figure 9)

S. magellanicum Brid.

(figure 7 and plates 2, 22, 27, and 28)

Plants large and robust, usually forming mats and low hummocks in open bogs, less common in shaded habitats. Plants in open areas are usually red to purple-red, sometimes pink, green in shaded habitats. Capitulum composed of short, swollen branches. Stem is red, outer layer of cells is easily removed. Branches arranged in fascicles of 2 + 2 (2 + 3); spreading branches swollen and tapering to tip (1.0–1.5 cm long); pendent branches long, tapering, and appressed to stem. Branch leaves large (1.5–2.0 mm long), ovate, and very concave, small teeth present near apex on back side of leaf. Stem leaves long and rectangular (up to 2 mm long), apex broadly rounded or minutely fringed.

Figure 7. *Sphagnum magellanicum* stem leaf (left) and branch leaf (right).

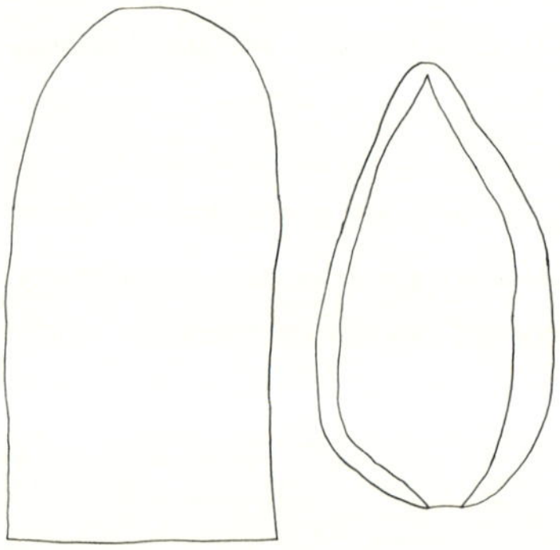

This is the only red-colored species in the section, but green plants are also common, especially in shaded habitats. Besides color, the most distinctive features of this species are the long rectangular stem leaves with nearly parallel sides. *Sphagnum capillifolium* var. *tenellum*, another common species of open bog habitats that is often red and is in section *Acutifolia*, is often mixed with this species, but the two are easily separated because *S. magellanicum* is a much larger plant and has cucullate branch leaves.

S. centrale C. Jens.
(figure 8 and plate 3)

Plants large and robust, forming carpets in shaded habitats often surrounding open peatlands. Plants are pale green to green. Capitulum composed of short, stout branches. Stem is green to light brown, outer layer of stem easily removed. Branches arranged in fascicles of 2 + 3 (2 + 2 or 2 + 4); spreading branches longer and more-or-less covering the stem. Branch leaves very large (1.5–2.5 mm long), ovate, and very concave, small teeth present near apex on back side of leaf. Stem leaves fan-shaped (usually 1.5-mm long or less), widest above the middle, broadly rounded apex with fine fringe sometimes extending down sides of leaf.

Very similar to *S. magellanicum*, but is primarily found in woodland habitats. The two species are often growing together, but *S. centrale* is distinguishable by its fan-shaped leaves and its lack of any red color.

Sphagnum palustre is similar looking but can only be reliably identified through a microscope, especially when the plants are green. See the discussion of *S. palustre* and other species in this section under *S. papillosum*.

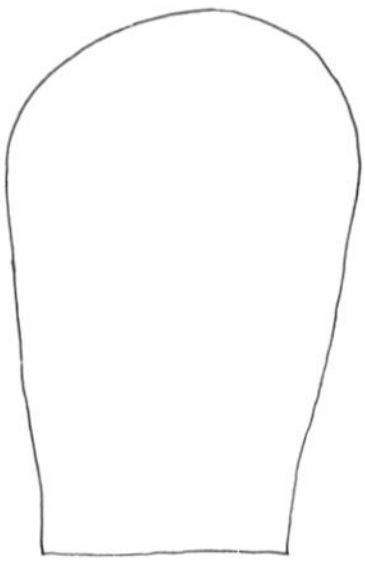

Figure 8. *Sphagnum centrale* stem leaf.

S. papillosum Lindb.

(figure 9 and plate 4)

Plants medium to large in size, forming floating mats and carpets in poor fens, rarely in shaded habitats. Plants tinged with yellow, often golden brown, yellow-brown, dirty brown, sometimes light orange, never red. Capitulum is very small, composed of very short, blunt branches. Stem is dark brown to almost black, covered by white layer of cells that is easily scraped off. Branches arranged in fascicles of 2 + 2, spreading branches blunt and swollen, pendent branches longer, almost covering stem. Branch leaves loosely arranged on branches, large (up to 2 mm long), ovate, or deep spoon-shaped. Stem leaves rarely longer than 1.5 mm, widest at or below middle, apex broad and rounded with fine fringe extending down the sides of the leaf.

Primarily a golden brown species, rarely found in woodland habitats. The branches are extremely short and blunt compared to other species in this section.

There are at least three other species that are barely distinguishable from *S. papillosum* in the field. These are *S. austinii*, *S. palustre*, and *S.*

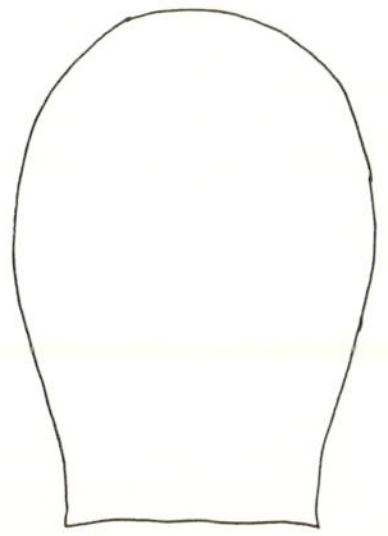

Figure 9. *Sphagnum papillosum* stem leaf.

henryense. *Sphagnum austinii* (plate 29) occurs in sedge fens, *Salix-Alnus* carrs, and pond margins where it forms mats and low hummocks. I have seen it in weakly minerotrophic peatlands along the coast of Maine, where it is commonly intermixed with *S. fuscum* and forms some large hummocks. The plants are described as being red-brown with a speckled appearance. In the field, I recognize it by its rich deep brown color, growing in large hummocks in direct sunlight. When growing intermixed with *S. fuscum*, it is easily distinguishable by its large size. *Sphagnum palustre* (plate 30) is found in a wide range of minerotrophic habitats, such as at the margins of bogs, fens, and ponds. This species can also form carpets and hummocks in *Salix-Alnus* carrs. The plants may develop squarrose leaves in shaded habitats and look very similar to *S. squarrosum*. They are green to golden brown, with long stem leaves (length nearly twice their width, 1.5 mm long or longer). *Sphagnum henryense* grows in forested fens, poor fens, and along pond margins. These plants have a reddish brown color, the newest branches of the capitulum are noticeably shorter than the branches below, there is a small but noticeable apical bud in the flat capitulum, and there are slightly squarrose leaves.

Section III. *Polyclada* (monotypic)

S. wulfianum Girg.

(figure 10 and plate 5)

Plants large and robust, very stiff, growing in loose carpets in woodland habitats on damp or fairly dry humus; plants not packed together. Plants dull green, brownish green, sometimes tinged with pink or light red in upper portions. Capitulum is very large, hemispherical, appearing shaggy, resembling a head of clover. Stem is dark brown to almost black, very stiff and thick (more than 0.5 mm thick). Individual fascicles have six to twelve branches with three to six spreading branches (6 + 5, 4 + 7, 4 + 4 common). Branch leaves are long (1.0–1.2 mm), narrowly ovate-lanceolate, with recurved tips, and arranged in five rows on branches. Stem leaves are small (not more than 0.8-mm long), triangular-lingulate, apex pointed or slightly rounded; may be fringed at apex.

Figure 10. *Sphagnum wulfianum* stem leaf (left) and branch leaf (right).

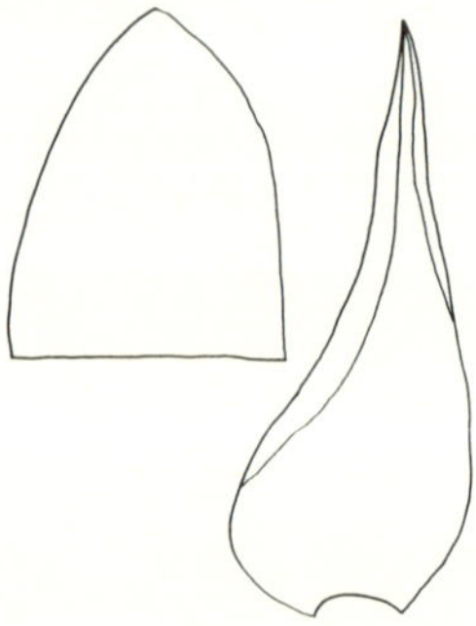

This species is easily recognized because it is the only peat moss with six or more branches per fascicle. Some species such as *S. angustifolium* and *S. quinquefarium* may look similar, but they have five or fewer branches per fascicle. The plants are not packed together like other species. The thick, dark stems are almost woody to the touch.

Section IV. *Squarrosa*

Species in this section typically have squarrose branch leaves. Some species in sections *Sphagnum* and *Rigida* also have somewhat squarrose branch leaves, but these species are easily distinguishable in the field. The species in section *Sphagnum* have cucullate branch leaves, whereas those in this section do not. The squarrose forms of species in section *Rigida* have very small stem leaves (less than 0.7 mm long), whereas the stem leaves of species in this section are much larger (more than 1.5 mm long). Thus, the key field characteristics are noncucullate, squarrose branch leaves and large stem leaves. Both species are common in mineral-rich habitats.

Key to Species

1. Branch leaves strongly squarrose throughout plant; stem leaves long triangular with truncated apex; pale green, rarely dark brown—*S. squarrosum* (figure 11 and plate 6)
1. Branch leaves imbricate in upper portion of plant and slightly squarrose in lower portion; stem leaves lingulate with rounded or flat apex, apex may have three to four low, broad teeth; plants brown in open areas, pale green in shaded areas—*S. teres* (figure 12 and plates 7 and 8)

S. squarrosum Crome
(figure 11 and plate 6)

Plants tall, robust, and stiff, forming loose carpets in shaded habitats, especially in coniferous forests; also common in rich fens among willows and alders; prefers moist sites, seldom found near water level. Plants pale to bright green. Prominent terminal bud present, surrounded by very short, loosely packed branches. Stem is pale green to red-brown. Branches arranged in fascicles of 2 + 2 (3 + 2), pendent branches closely covering the stem. Branch leaves large (0.9–2.5 mm long), definitely squarrose, with apex bent from 45 to 90 degrees from branch axis; lower half of leaf nearly oval, tapering sharply to tip. Stem leaves large (1.5–2.5 mm long), lingulate with broad, rounded apex, often fringed at tip.

Figure 11. *Sphagnum squarrosum* stem leaf (left) and branch leaf (right).

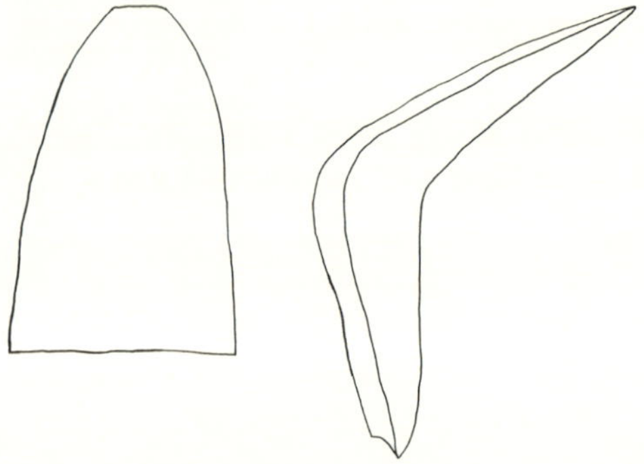

The strongly squarrose branch leaves make this species easy to recognize in the field. Typical forms are green, but yellowish brown forms are common and may be confused with *S. teres*. *S. squarrosum*, however, typically has two spreading and two pendent branches per fascicle, whereas *S. teres* frequently has three spreading branches per fascicle and is not common in shaded habitats.

S. teres (Schimp.) Angstr.

(figure 12 and plates 7 and 8)

Plants medium-sized and fairly slender, forming loose or dense carpets in open fens; seldom in shaded habitats. Plants yellowish, yellow brown to red-brown, especially in unshaded habitats. Terminal bud large; small stellate capitulum. Stem green in upper portion and red-brown in lower portion. Branches arranged in fascicles of 3 + 2 (2 + 3); spreading branches slender, extended nearly perpendicular to stem; pendent branches tightly appressed to stem. Branch leaves only slightly squarrose when moist and more strongly squarrose when dry, apex of leaf bent 30 to 60 degrees from branch axis, rarely 90 degrees, ovate to lingulate-lanceolate with tip abruptly narrowed; apex with small teeth. Stem leaves large (1.3–2.4 mm long), oblong-lingulate to spatulate, widest at or below middle, apex broad-rounded and fringed across top.

Figure 12. *Sphagnum teres* stem leaf (left) and branch leaf viewed from top (middle) and side (right).

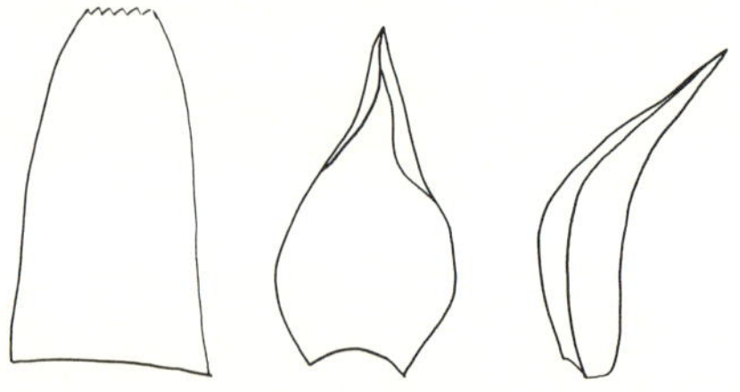

Reliable field characteristics are the rather rigid branches nearly perpendicular to the stem, slightly squarrose branch leaves when dry, yellowish to brown color, and large terminal bud. Superficially, green forms of this species may resemble *S. girgensohnii*, but they differ in having a dark stem, three spreading branches per fascicle, and a tolerance for sunlight. Both *S. squarrosum* and *S. teres* are common and frequently grow in very minerotrophic habitats.

Section V. *Subsecunda*

Five species and a great number of varieties of this section grow in mineral-rich habitats. Most of the diagnostic characteristics are microscopic; however, the three species that are the most likely to be encountered are easily distinguishable using macroscopic features.

Key to Species

1. Plants slender, unbranched, or with one or two branches per fascicle; capitulum absent—*S. pylaesii* (figure 13 and plate 9)
1. Plants large, with branches arranged in fascicles of three or more branches; capitulum present
    2. Stem leaf margin flat, only slightly longer than broad, rounded at apex; branch leaves not much longer than stem leaves, margins inrolled nearly entire length, leaves slightly curved to one side; small, slender, wiry plants; capitulum usually without a terminal bud; stem is dark brown—*S. subsecundum* (figure 14 and plate 10)
    2. Stem leaf margin flat or inrolled, nearly twice as long as broad, slightly rounded or nearly acute at apex; branch leaves nearly twice the length of the stem leaves, inrolled only near the apex, slightly curved to one side; capitulum large, flat, terminal bud present; stem is pale green to light brown—*S. contortum* (figure 15 and plate 11)

S. pylaesii Brid.

(figure 13 and plate 9)

Plants slender, sparsely branched; unbranched forms common in aquatic habitats; prostrate when growing over wet rocks, forming small tufts when growing in poor fens. Plants dark green to purple-brown in submerged forms, pinkish to salmon-red in exposed forms. Terminal bud large, at most only a few branches are clustered around the terminal bud. Stem is pale green, yellow to brown in the lower portions of the plant. Branches lacking or arranged in fascicles of 2 + 0 (2 + 1), branch leaves loosely arranged on branches. Branch leaves when present are similar to stem leaves only smaller (0.8–1.3 mm long), ovate, sometimes as long as wide; deeply concave, rounded apex. Stem leaves broadly ovate and concave, large in comparison with stem (1.4–2.5 mm long).

Unbranched forms of this unusual peat moss are common in seepage areas on granitic rock in mountainous areas, while aquatic forms in high mountain ponds may be sparsely branched. This form is considered to be a pioneer species.

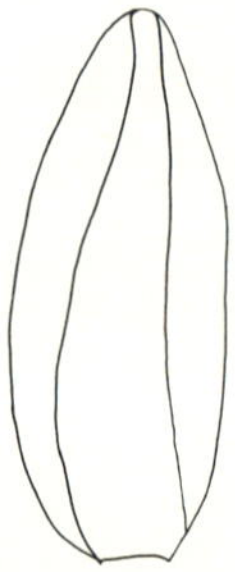

Figure 13. *Sphagnum pylaesii* stem leaf.

S. subsecundum Nees

(figure 14 and plate 10)

Plants small and slender, wiry, forming loose carpets in rich fens, wet sedge meadows, and on margins of ponds; intolerant of shade. Plants green, yellow, orange, to dark brown. Terminal bud small; capitulum composed of small curled branches. Stem is yellow-green or light brown in upper portion to dark brown in lower portion. Branches arranged in fascicles of 2 + 3, 2 + 4, 3 + 3; spreading and pendent branches similar in appearance and length (usually less than 1 cm long); pendent branches appressed to stem. Branch leaves nearly twice as long as stem leaves (up to 1.2 mm long or longer), ovate to ovate-lanceolate, deeply concave, curved to one side and slightly secund, loosely arranged on branch when moist, appressed to branch when dry. Stem leaves small (less than 0.8 mm long), width nearly the same as length, triangular-lingulate, lingulate, concave; broad-rounded apex, sometimes with small teeth.

Dry plants appear very shiny. The combination of yellow and orange color, dark stems, curved branches in capitulum, curved branch leaves, and mineral-rich habitat make this species easy to recognize in the field. Two closely related species, *S. lescurii* and *S. platyphyllum*, are larger, but they

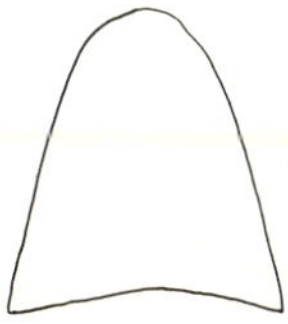

Figure 14. *Sphagnum subsecundum* stem leaf.

are extremely variable and difficult to distinguish in the field from *S. subsecundum.* These two species are included in the key to nearly all species at the end of the book. The form may be confused with *S. contortum,* but that species is larger in size and not stiff or wiry.

Sphagnum tenellum, a species in section *Cuspidata,* may sometimes be confused with this species; however, *S. tenellum* grows in nutrient-poor habitats such as wet hollows, pools in ombrotrophic bogs, poor fens, and on wet rocks at high elevations.

S. contortum Schultz

(figure 15 and plate 11)

Plants moderate-sized, weak-stemmed, appearing very floppy; forms loose cushions and carpets in rich minerotrophic habitats; intolerant of shade. Plants pale green to light brown. Capitulum is large and flattened with small terminal bud, short branches of capitulum tightly curled with most of them pointing upwards. Stem is pale green in upper portion to light brown in lower portion. Branches arranged in fascicles of 2 + 2, 2 + 3, 3 + 3, (2 + 1); spreading branches clearly different from pendent branches loosely covering stem. Branch leaves ovate to ovate-lanceolate (1.0–2.0-mm long), concave, curved, noticeably secund, and pointing towards center of curve of branch; leaves loosely arranged on branch. Stem leaves triangular to triangular-lingulate (0.7–1.3 mm long); apex rounded.

This is perhaps the most minerotrophic of all peat mosses, along with *S. teres* and *S. warnstorfii*. It is intolerant of shade and grows in nearly basic habitats. Plants are typically weak-stemmed and grow in loose, tangled masses. Plants are often a light yellow to light golden brown with a pale green or light brown stem, which should serve to distinguish it from *S. subsecundum*. Often found with *S. teres*.

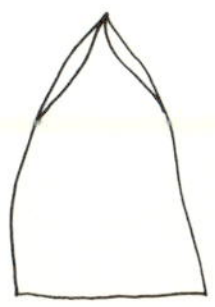

Figure 15. *Sphagnum contortum* stem leaf.

Section VI. *Acutifolia*

The species in this section are common in acid bog habitats. Most form compact cushions or hummocks, and may form extensive mats in the drier portions of bogs. Color is useful to group the species in this section. Red species include *S. capillifolium, S. subtile, S. warnstorfii, S. russowii,* and *S. quinquefarium. Sphagnum fuscum* is the only common brown species, and *Sphagnum girgensohnii* and *S. fimbriatum* are typically green. A distinctive feature of the forms with stellate capitula is the absence or the presence of only a single branch between each of the five rays of branches. This feature is useful for separating these species from the species of section *Cuspidata,* which have two branches between rays. In general, these plants grow in more acidic habitats than the other species of peat moss.

Key to Species

1. Many fascicles with three spreading branches and pendent branches—*S. quinquefarium* (figure 16 and plate 12)
1. Two spreading branches and one or two pendent branches per fascicle
    2. Stem leaves fan-shaped or lingulate and fringed at least at apex
        3. Stem leaves fan-shaped and fringed from apex nearly to base; terminal bud large; often fruiting—*S. fimbriatum* (figure 17 and plates 13, 14, and 20)
        3. Stem leaves lingulate and fringed only at apex or with distinct notch; terminal bud

lacking or poorly developed; seldom fruiting

4. Stem leaves fringed across apex; plants stiff and wiry with long, slender spreading branches perpendicular to stem; plants green, never red—*S. girgensohnii* (figure 18 and plates 3, 15, and 16)
4. Stem leaves with distinct notch at apex; plants not especially wiry or stiff; plants green, but often splotched with red—*S. russowii* (figure 19 and plates 3 and 16)

2. Stem leaves variously shaped, entire margins, never fimbriate, with rounded or pointed apex
   5. Flat, five-part capitulum; branch leaves in five rows
      6. In swamps with plants of *Thuja*, *Abies*, and/or *Picea*, and in minerotrophic fens; slender spreading branches with leaves in five distinct rows, leaves straight not curved; stem leaves short, with a ratio of length to width of 1.5:1 or less, lingulate or triangular with apex usually rounded, but may be pointed; plants often purplish red throughout, but may be green in shaded habitats—*S. warnstorfii* (figure 20 and plate 17)
      6. In open bogs, commonly forms hummocks; branch leaves usually in five rows, leaves usually curved to one side; stem leaves long lingulate, with a length to width ratio of 1.5:1 or longer, with rounded or pointed apex; plants green or red throughout, or splotched red-green—*S. capillifolium* (var. *tenellum*) (figure 22 and plates 18, 27, and 28)

5. Capitulum flat, rounded, or hemispherical, not five-parted; branch leaves not in five rows
   7. Plants brown, greenish brown, or yellowish brown; capitulum flat; stem leaves lingulate with broad, rounded apex; forms hummocks in dry, open, acid bogs—*S. fuscum* (figure 23 and plates 19, 28, and 29)
   7. Plants not brown, usually green, splotched red and green, or red; capitulum rounded or hemispherical; stem leaves lingulate or triangular with pointed apex
      8. Stem leaves short, usually not much longer than 1 mm, length to width ratio less than 1.5:1, triangular; capitulum rounded; forms hummocks on damp, coniferous humus, sometimes in hardwood forests—*S. subtile* (figure 24 and plate 16)
      8. Stem leaves long, usually longer than 1 mm, length to width ratio greater than 1.5:1, lingulate, somewhat ovate, not triangular; capitulum on small plants rounded, on large plants, hemispherical; forms hummocks in acid bogs with and without trees—*S. capillifolium* (var. *capillifolium*) (figure 21 and plate 18)

S. quinquefarium (Braithw.) Warnst.

(figure 16 and plate 12)

Plants slender, tall, densely branched, forming mats on sheltered seepage areas, wet dripping cliffs, sloping banks in mountains, and on peaty soil in swamps. Plants are light green to dark green, sometimes tinged with pink or red. Capitulum is well developed, round to hemispherical. Stem is pale green to yellow, nearly white in the upper portion. Branches arranged in fascicles of 3 + 2 (2 + 3); spreading branches slender and gracefully curved (some may be straight), branch leaves arranged in five distinct rows on each spreading branch; pendent branches loosely cover stem (stem is largely hidden due to the great number of branches). Branch leaves are ovate to ovate-lanceolate (0.8–1.6 mm long). Stem leaves triangular to triangular-lingulate, nearly as long as broad (1.0–1.4 mm long); apex rounded or acute, sometimes involute.

Figure 16. *Sphagnum quinquefarium* stem leaf (left) and 3 + 2 fascicle (right).

Superficially, this species resembles *S. subtile* because of its small, triangular stem leaves, nearly hemispherical capitulum, and habitat. This species is the only one in this section, however, to have three spreading branches per fascicle, and the branch leaves are arranged in five distinct rows on the branches.

S. fimbriatum Wils.

(figure 17 and plates 13, 14, and 20)

Plants very slender, but very elongate, growing in very loose tangled carpets among sedges in open fens and at the margins of bogs. Color is pale green, gray-green, or pale yellow. Capitulum is small with a conspicuous terminal bud. Stem is thin (less than 0.5 mm) and pale green to yellow. Branches arranged in fascicles of 2 + 2 (2 + 1), branches very long and thin; spreading branches gracefully curved; pendent branches loosely covering stem. Branch leaves small to medium-sized (0.8–2.0 mm long), ovate to ovate-lanceolate, strongly concave and involute near apex. Stem leaves are fan-shaped or spoon-shaped, clasping the stem, often broader than long (0.8–1.3 mm long), fringed and tattered across the broad, rounded apex and down the sides.

No other species produces sporophytes more freely than this one. In early summer the capitula are crowded with capsules. This species is one of

Figure 17. *Sphagnum fimbriatum* stem leaf.

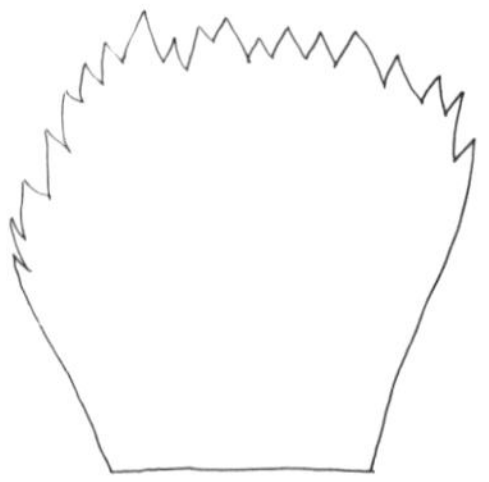

the most minerotrophic of this section. Its most distinctive feature is the fan-shaped, fimbriate stem leaves. Stem leaves appear as a "ruff" around the stem, forming a shallow bowl on decapitated plants.

Only one other species has fan-shaped leaves—*S. lindbergii*—but this species can scarcely be confused with *S. fimbriatum*. *Sphagnum lindbergii* grows in thick carpets in ombrotrophic to poor fen habitats. Unlike *S. fimbriatum*, *S. lindbergii* has a dark stem and five-ranked branch leaves.

S. girgensohnii Russow
(figure 18 and plates 3, 15, and 16)

Plants large, stiff, and wiry, forming loose carpets on damp humus in woodland habitats; common in woodlands surrounding bogs. Plants are green, never red. Capitulum is small, flat, and stellate, with small terminal bud. Stem is pale green, green, or pale yellow, very stiff, breaks crisply when snapped in two. Branches arranged in fascicles of 2 + 2 (2 + 1); spreading branches nearly perpendicular to stem, long and drooping, curved downward near tip; pendent branches loosely covering stem. Branch leaves ovate-lanceolate to nearly lanceolate (1.0–1.4 mm long), margins of leaves strongly involute near apex. Stem leaves lingulate, a little longer than broad (1.0–1.3 mm long), leaves appressed tightly to stem, apex truncate and coarsely fringed (dentate).

This species is one of the most common of the peat mosses. Distinctive features are its flat stellate capitulum, spreading branches growing nearly

Figure 18. *Sphagnum girgensohnii* stem leaf.

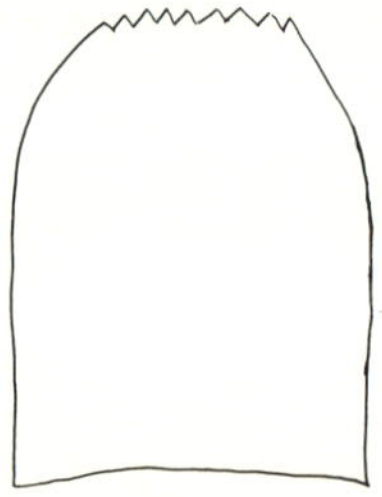

perpendicular to the stem, and the nearly square stem leaves with a broad fimbriate (dentate) apex. The plants are extremely stiff and wiry. Snapping the stem in two reminds one of the crisp sound of breaking fresh celery. *Sphagnum russowii* often grows intermixed with *S. girgensohnii*, but the two are easily separated on the basis of color, since *S. russowii* is usually splotched with red whereas *S. girgensohnii* lacks any red pigment.

S. russowii Warnst.

(figure 19 and plates 3 and 16)

Plants large, forming small hummocks and loose carpets on damp humus in woodland habitats, in shaded habitats of open bogs, and in wet coniferous forests in mountains; frequently mixed with other species. Plants are multicolored—green splotched with red, pale pink in unshaded habitats. Capitulum is flat and stellate; branches of capitulum are often club-shaped. Stem is green or tinged with red. Branches arranged in fascicles of 2 + 2 (2 + 1), branches in upper portion usually less than 1 cm long, branches in lower portion frequently greater than 1 cm long. Spreading branches are slender and gracefully curved from the base; pendent branches loosely cover the stem. Branch leaves are ovate to ovate-lanceolate, strongly involute at apex. Stem leaves are lingulate (0.9–1.4 mm long), apex broadly rounded with small notch in the center.

Figure 19. *Sphagnum russowii* stem leaf.

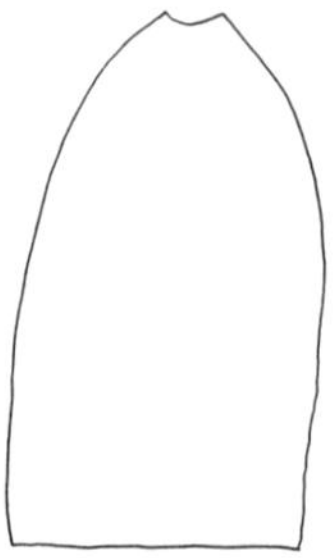

This species is seldom abundant but does grow in a large number of habitats. Plants growing in shaded habitats look similar to *S. girgensohnii,* but the plants are typically splotched red and not stiff or wiry. The stem leaves are not as square as those of *S. girgensohnii* and typically have a small notch at the apex. This feature is also useful in distinguishing *S. russowii* from *S. capillifolium. Sphagnum russowii* has a lighter red to almost pinkish color in open growth forms that does not occur in other species.

S. warnstorfii Rossow

(figure 20 and plate 17)

Plants small and slender, forming small, loose hummocks or carpets on wet humus in rich, shaded, minerotrophic habitats such as cedar bogs. Plants multicolored, green splotched with dark red, purple-red, rarely completely red or green throughout. Capitulum flat and stellate. Stem is red or green in upper portions to almost brown in lower portions. Branches arranged in fascicles of 2 + 2 (2 + 1); spreading branches slender and strongly divergent from stem, branch leaves arranged in five distinct rows; pendent branches loosely covers the stem. Branch leaves are small (0.6–1.4 mm long), ovate to ovate-lanceolate, recurved near the apex on dry branches (usually much less than 90 degrees), straight on moist branches. Stem leaves are lingulate with broad, rounded apex (0.8–1.3 mm long) and a shallow notch.

Figure 20. *Sphagnum warnstorfii* stem leaf.

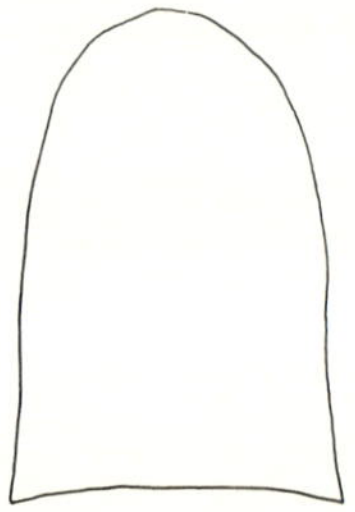

A common mat-forming species in cedar swamps and rich fens. These plants have a purplish red color that is not found in other species of this section. In shaded habitats, the plants may also be dark green. *S. capillifolium* var. *tenellum* can be confused with this species, although *S. capillifolium* var. *tenellum* has long, lingulate stem leaves with a narrower apex, has branch leaves that are purplish red, and is not tolerant of shaded, minerotrophic habitats.

S. fuscum (Schimp.) Klinggr.

(figure 21 and plates 19, 28, and 29)

Plants small and slender forming very large, compact, dry hummocks in the drier, open portions of acid bogs. Plants are greenish at the top, but mostly green-brown to brown. Capitulum is small, flat, and stellate. Stem is small (0.2–0.5 mm), green in upper portion, brown in lower portion. Branches arranged in fascicles of 2 + 2 (2 + 1), spreading branches short (up to 1-cm long), pendent branches tightly appressed to stem and nearly twice as long as spreading branches. Branch leaves are small, ovate-lanceolate (0.9–1.3 mm long), strongly involute at apex. Stem leaves lingulate (0.8–1.3 mm long); the broadly rounded apex smooth or with some very small teeth.

The deep brown to greenish brown color of these plants make them easy to recognize. *S. fuscum* grows at the tops of the larger hummocks in one of the driest (usually damp, but dry for peatland),

Figure 21. *Sphagnum fuscum* stem leaf.

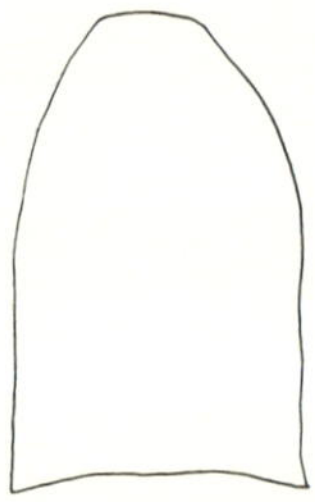

most acidic habitats of any peat moss. The red-colored varieties of *S. capillifolium* often inhabit on the sides and bases of these hummocks, with pale-colored species such as *S. fallax* and *S. majus* occurring in the wet hollows. Vascular plants such as Labrador tea, sheep laurel, cranberry, and blueberry are common on these hummocks.

Other brown-colored species that may be confused with *S. fuscum* are *S. flavicomans*, *S. subfulvum*, and *S. subnitens*. In general, these plants are larger, including the stem leaves that are 1.5 mm or more in length. *Sphagnum flavicomans* is a much larger, looser plant than *S. fuscum* and forms hummocks along the margins of ponds and in the shaded portions of poor fens, habitats quite different from those preferred by *S. fuscum*. The stem leaves of *S. flavicomans* are also different, in that they are narrow lingulate-triangular in shape, and the capitulum is quite often large and hemispherical. *Sphagnum subfulvum* grows in much more minerotrophic habitats than *S. fuscum*, such as medium and rich fens in the open or under shrubs. It also grows closer to water level than *S. fuscum*. The capitulum of *S. subfulvum* is large and flattened, and the stem leaves are triangular-lingulate with a rounded or acute apex. *Sphagnum subnitens* is found in ombrotrophic bogs and poor fens. *Sphagnum subnitens* is similar to *S. subfulvum*, but it has triangular-lingulate leaves with an abruptly narrowed, cuspidate apex, and it has tinges of red and a shiny appearance when dry.

S. subtile (Russow) Warnst.

(figure 22 and plate 16)

Plants small, compact, and stiff, forming small, compact hummocks on damp humus in woodland habitats; especially common in coniferous forests and woodlands surrounding bogs. Plants are green, tinged with red in the upper portion of the plant. Capitulum is round to nearly hemispherical. Stem is green to yellowish green, relatively stiff. Branches arranged in fascicles of 2 + 2 (rarely 2 + 1), densely crowded, long and slender (up to 1.5 cm long), spreading branches loose and tangled or recurved, pendent branches very slender and appressed to stem. Branch leaves are ovate to ovate-lanceolate (1.0–1.5 mm long). Stem leaves are small (0.8–1.1 mm long), triangular to triangular-lingulate, apex acute.

This plant closely resembles *S. capillifolium* (especially var. *capillifolium*) but is easily distinguished by its small, triangular-shaped stem leaves and typical woodland habitat. Most plants are green, although those growing near openings in the forest canopy may contain some red.

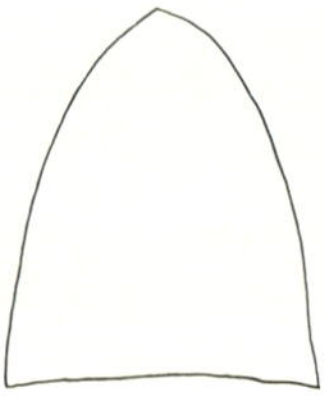

Figure 22. *Sphagnum subtile* stem leaf.

***S. capillifolium* var. *capillifolium* (Ehrh.) Hedw.**

(figure 23 and plate 18)

Plants small, compact, and stiff, forming compact, dry hummocks in the drier portions of open acid bogs and on the base and sides of *S. fuscum* hummocks. Plants are usually pale green to green tinged with red or red throughout. Capitulum is round to hemispherical. Stem is pale green in upper portion, red in lower portion. Branches arranged in fascicles of 2 + 2 (2 + 1), spreading branches long (up to 1.5 cm), loose, and tangled or recurved, pendent branches very slender, tightly appressed to stem. Branch leaves are appressed to branch, ovate-lanceolate to lanceolate (1.0–1.8 mm long), slightly involute near apex. Stem leaves are lingulate to lingulate-triangular (1.0–1.8 mm long), apex definitely involute.

The species is typically found in the drier portions of open acid bogs. It often dominates some of the larger hummocks and the base and sides of *S. fuscum* hummocks. These red-colored plants are often confused with *S. capillifolium* var. *tenellum*,

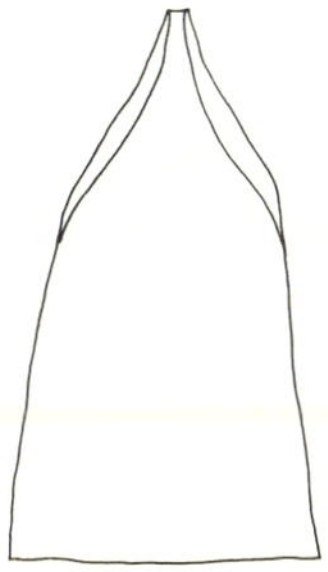

Figure 23. *Sphagnum capillifolium* var. *capillifolium* stem leaf.

but unlike the latter variety, *S. capillifolium* var. *capillifolium* has a large, hemispherical (pom-pom like) capitulum, pointed stem leaves involute at the tip, and straight branch leaves without any apparent organization of leaves into rows on the branches. This variety also typically bears more sporophytes than var. *tenellum*. Both varieties are often found together at the base and sides of larger hummocks. Although the varieties are quite difficult to distinguish from one another, there is apparently a genetic basis for their separate classification.

S. capillifolium var. *tenellum* (Schimp.) Crum

(figure 24 and plates 18, 27, and 28)

Plants slender, forming extensive, dense carpets and hummocks in open acid bogs. The color is generally red or green splotched with red. Plants are intolerant of shade. Capitulum is flat and stellate but may be rounded when growing on hummocks. Stem is green in upper portion to red in lower portion. Branches arranged in fascicles of 2 + 2 (2 + 1). Spreading branches slender, strongly divergent from stem, with branch leaves in five rows (most noticeable on dry branches). Pendent branches appressed to stem. Branch leaves are ovate to ovate-lanceolate, slightly curved, and somewhat secund. Stem leaves are large (up to 1.5 mm long), lingulate to lingulate-triangular, apex rounded or acute; may be involute near apex.

One of the most common mat-forming peat mosses in open acid bogs, these plants are often found growing on floating mats and forming low hummocks at the base and sides of larger hummocks dominated by *S. fuscum*. Most plants grow-

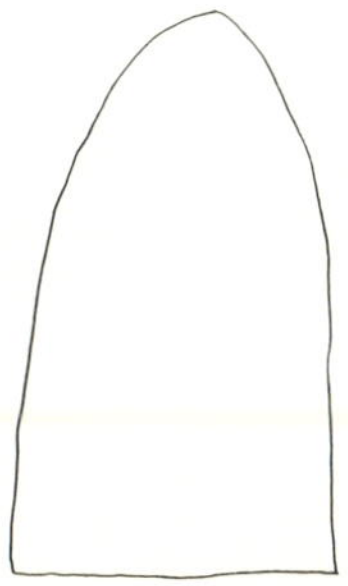

Figure 24. *Sphagnum capillifolium* var. *tenellum* stem leaf.

ing in direct sunlight are brilliant red. In wetter habitats, this variety is very distinct from *S. capillifolium* var. *capillifolium*; however, in drier, more ombrotrophic peatlands where both varieties are common, it is virtually impossible to distinguish them. In such cases, the plants should simply be called *S. capillifolium*.

Sphagnum bartlettianum (plate 30) is macroscopically virtually identical to *S. capillifolium* var. *tenellum*, however their ranges and habitats are different. *Sphagnum bartlettianum* is more common in the southeastern United States, extending up the east coast to New England. It is usually found in richer sites than is *S. capillifolium* var. *tenellum* and occurs with species not normally associated with *S. capillifolium* var. *tenellum*. Its typical habitat is on sandy soil in pine barrens, shrubby swamps, and medium to rich fens, habitats that are not common in the northern United States.

Section VII. *Cuspidata*

In the *Sphagnum* hummock-hollow sequence, the species in this section occupy the wetter hollows and pools, except for *S. angustifolium*, which may be found from near the tops of *S. fuscum* hummocks to the wet hollows. The species with a five-part or stellate capitulum have paired branches between the rays of the branches. This feature, along with the absence of red coloration, is useful in distinguishing these plants from those of section *Acutifolia* with which they are commonly found. In general, *Cuspidata* species grow in the wetter habitats of open acid bogs and may be the dominant peat mosses in minerotrophic habitats such as sedge fens. Most species grow in very loose mats and are very weak stemmed. Capitula are easily detached, and care must be taken in plucking off fascicles to examine the stem leaves. The plants often feel and look like overcooked vegetables. A distinctive fish odor is noticeable even from dried specimens.

Key to Species

1. Stem leaves large, 1–1.5 mm long, apex irregularly torn down the middle—*S. riparium* (figure 25 and plate 20)
1. Stem leaves large and small, apex not cut or torn down the middle
    2. Branch leaves long, most often over four times longer than the width
        3. Stem leaves with acute apex; branch leaf margins inrolled from middle and below to

apex; plants green or yellow, bases of branches in capitulum may be tinged with red; often grows submerged in pools and wet depressions and along lake margins—*S. cuspidatum* (figure 26 and plates 21 and 22)

3. Stem leaves with rounded apex; branch leaf margins inrolled only near apex; plants brownish green to almost black; sometimes grows submerged, but most often near water level in wet depressions—*S. majus* (figure 27 and plate 22)

2. Branch leaves short, seldom reaching four times longer than width
   4. Stem leaf apex truncated or rounded
      5. Stem leaves small, less than 0.8 mm in length; stem near apex often light pink; pendent branches loosely cover stem—*S. angustifolium* (figure 28 and plates 3, 24, 27, and 28)
      5. Stem leaves greater than 0.8 mm in length; stem near top of plant is light green, rarely pink; pendent branches mostly cover stem
         6. Branch leaves in five rows; pendent and spreading branches and leaves similar, with pendent branches loosely surrounding the stem; branch leaves long and narrowly ovate-lanceolate; capitulum five-parted—*S. recurvum* (figure 29 and plates 23, 24, and 30)
         6. Branch leaves seldom in five rows; pendent and spreading branches different, with pendent branches tightly

covering the stem; branch leaves ovate to broadly ovate-lanceolate; capitulum resembles a small, loose ball of yarn—*S. flexuosum* (figure 30 and plates 23, 24, and 30)

4. Stem leaf apex acute or apiculate
   7. Branches short and thick with broad apiculate leaves, branch leaves in five rows; spreading and pendent branch leaves different; stem leaves with acute apex; plants green, yellow, or brown—*S. pulchrum* (figure 31 and plate 25)
   7. Spreading and pendent branches slender, branch leaves not in five distinct rows (branch leaves may be arranged in five rows but not as neatly as *S. pulchrum*); spreading and pendent branch leaves similar; stem leaves with sharp apiculate apex; plants green, yellow, seldom brown—*S. fallax* (figure 32 and plate 26)

S. riparium Angstr.
(figure 25 and plate 20)

Plants tall and rather stiff, forming extensive, loose carpets in wet minerotrophic habitats; rarely submerged. Plants are pale green or yellow, shiny when moist. The terminal bud is very distinct. Stem is very stout (0.5–1.0 mm thick), pale green to green. Branches arranged in fascicles of 2 + 2 (2 + 3), spreading branches stout, swollen, and pointed at the tips, pendent branches slender and appressed to stem. Branch leaves are large (1.3–2.5 mm long), ovate-lanceolate, recurved at tips when dry, concave when moist. Stem leaves are triangular-lingulate (1.3–1.6 mm long), with a conspicuous tear down the middle.

This species is easily recognized in the field but is often overlooked as just another carpet of *S. girgensohnii.* Its most distinctive features are its large terminal bud, stem leaves with a tear down the middle, and swollen, stout branches with pointed tips. This species is often found with *S. angustifolium* and *S. fallax,* and infrequently with *S. recurvum,* which is found from Nova Scotia inland to the Great Lakes and south throughout the eastern United States. *Sphagnum riparium* has a more northern distribution, ranging from the boreal north to the northern portions of the United States.

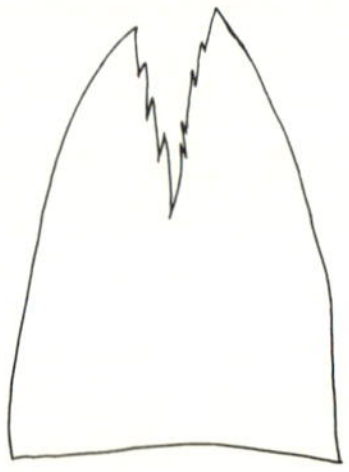

Figure 25. *Sphagnum riparium* stem leaf.

S. cuspidatum Hoffm.

(figure 26 and plates 21 and 22)

Plants large, slender, with a wet feathery appearance; forms very loose, tangled masses on wet peat and is found submerged in pools or often among sedges in minerotrophic habitats. Plants are bright green to yellow; the branches in the capitulum may have slight reddish tinge at the base. Capitulum is not well developed nor very noticeable, and the terminal bud is small. Stem is yellow-green. Branches arranged in fascicles of 2 + 2, although there is hardly any difference between spreading and pendent branches, which can be straight or curved. Pendent branches do not conceal the stem. Branch leaves are very long (1.5–3.0 mm or more), lanceolate, and curved, especially near the tips of the branches where sometimes almost all the leaves curve in the same direction. The branch leaves are involute from the apex almost to the middle. Stem leaves are smaller than branch leaves (1.0–1.3 mm

Figure 26. *Sphagnum cuspidatum* stem leaf (left) and branch leaf (right).

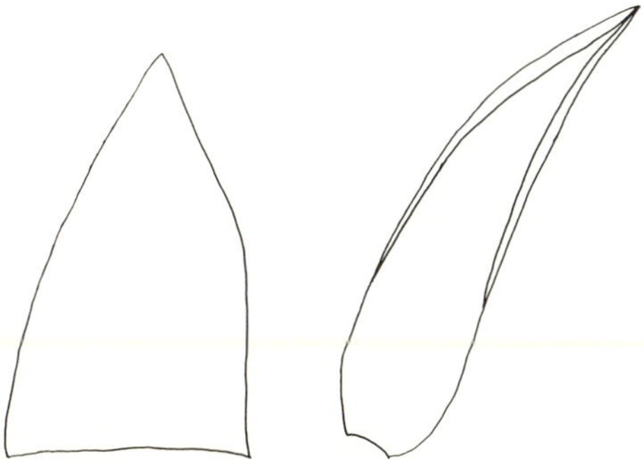

long), triangular, longer than they are broad, with the apex acute or slightly rounded with small teeth.

Primarily an aquatic species, these plants are often submerged or near the water level at the edges of floating mats and in shallow pools. Distinguishing features are its long, lanceolate branch leaves, sharp, pointed stem leaves, and very weak stems. Loosely tangled masses growing in pools are barely recognizable as peat mosses. Pendent branches are not well developed, which aids in distinguishing the species from *S. recurvum*. *Sphagnum majus*, a closely allied species that grows near the water level, is similar but has a dingy brown-green to blackish color, whereas *S. cuspidatum* is green or yellow.

Sphagnum torreyanum is larger and more robust than *S. cuspidatum* and also grows submerged or at water level. Macroscopic features helpful in distinguishing it are its branch leaves, inrolled from near the base to the apex, its stem leaves, whose sides and base are almost of equal length, and its prominent capitulum. Some plants cannot be distinguished from *S. cuspidatum*, however, without the use of a microscope.

S. majus (Russow) C. Jens.

(figure 27 and plate 22)

Plants large, sometimes growing submerged but most often forming loose carpets on wet peat near water level; intolerant of shade. Plants are dark-colored—green, brownish, yellow-green—but most commonly dirty green to almost black, especially in branch tips, as though scorched. The terminal bud is noticeable, the branches of the capitulum are blunter than the lower branches. Stem is yellow-green. Branches arranged in fascicles of 2 + 2 or 3 + 1, spreading branches usually longer than pendent branches and often curved and tapered, pendent branches loosely covering stem or slightly spreading. Branch leaves are large (up to 3.0 mm long), ovate-lanceolate, often curved and secund, involute in upper portion but not reaching the middle. Stem leaves are smaller than branch leaves (0.9–1.4 mm long), triangular-lingulate, with apex blunt or rounded, possibly with small teeth.

Figure 27. *Sphagnum majus* stem leaf (left) and branch leaf (right).

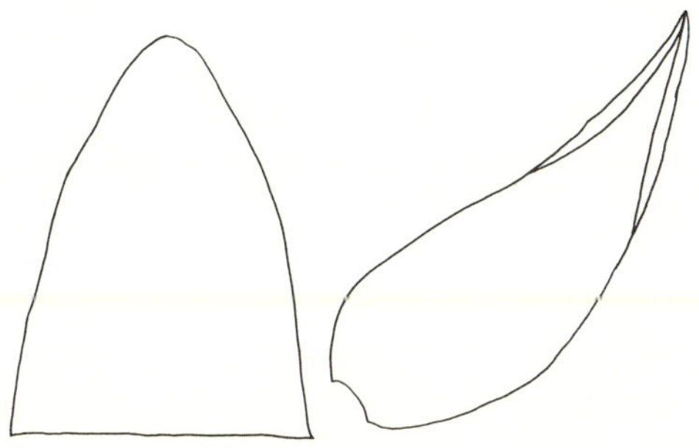

These plants grow in wet depressions that periodically dry out. They have a burned appearance, typically dingy green or brown in color, a useful distinction from *S. cuspidatum,* which is always green. Submerged forms resemble *S. cuspidatum,* and terrestrial forms resemble *S. recurvum,* but the dirty color of *S. majus* makes it easy to identify. This species is one of this section that lacks paired branches between the rays of the capitulum.

S. angustifolium (Russow) C. Jens.

(figure 28 and plates 3, 23, 24, 27, and 28)

Plants small- to moderate-sized and slender, forming dense carpets and small hummocks above water level in open acid bogs as well as in open and woodland fens. Plants are pale green to green, yellow-brown. Capitulum is small, flat or rounded, with bases of branches often tinged with pink. Stem is often pink, pale green, or yellow. Branches arranged in fascicles of 2 + 2, (2 + 3, 2 + 1), spreading branches usually shorter than pendent branches (up to 8 mm long), with leaves arranged in five rows on the branches, pendent branches threadlike and nearly white (up to 10 mm long), loosely covering stem. Branch leaves are small (up to 1.0 mm long), ovate-lanceolate, undulate, and slightly recurved on dry branches. Stem leaves are small (less than 0.8 mm long), triangular (nearly equilateral), with apex blunt, sometimes with small teeth.

This species has one of the broadest habitat ranges of any of the peat mosses. It is most abundant in depressions above water level in open bogs but can also be found on the sides of larger hummocks dominated by *S. fuscum*, *S. capillifolium*, or *S. magellanicum*. It is easily distinguished from these species by its pale green color, small stem

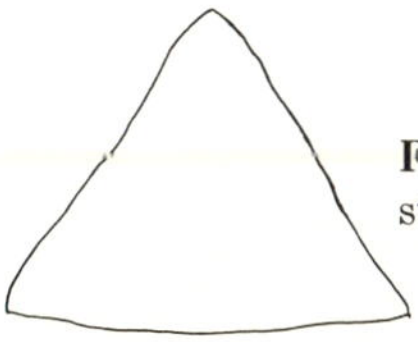

Figure 28. *Sphagnum angustifolium* stem leaf.

leaves, and the presence of paired branches between the rays of the capitulum. In wet depressions, it is often intermixed with *S. fallax* but is easily separated from that species by its small, blunt-tipped stem leaves, green color, and pink in the stem and bases of the upper branches. In woodland swamps, it can be found on damp humus and resembles *S. wulfianum* because of its large, dense capitulum that develops in this habitat. *S. angustifolium* has only two spreading branches per fascicle, however, whereas *S. wulfianum* has six or more branches per fascicle. The stem of *S. wulfianum* is also very dark, almost black, whereas the stem of *S. angustifolium* is green or pink, but never dark brown or black.

S. recurvum P. Beauv.

(figure 29 and plates 23, 24, and 30)

Plants fairly large, robust, and very soft looking; forms loose carpets close to water level in wet depressions of open bogs and at the edge of surrounding woodlands. Plants are pale green, green, to yellowish. Capitulum is large, densely packed with branches. Stem is pale green to yellow. Branches arranged in fascicles of 2 + 2 (2 + 3, 2 + 1); spreading branches straight, not strongly curved, leaves arranged in five rows on branches; pendent branches very slender, long, nearly white, tightly appressed to stem. Branch leaves are long (1.2–2.0 mm long), lanceolate, undulate, and slightly recurved when dry. Stem leaves are triangular, triangular-lingulate (0.8–1.3 mm long), with apex broadly rounded, truncated, and often appearing to be torn and ragged.

Shaded forms are pale green, while forms growing in direct sunlight may be yellow or yellowish brown. This species is very similar to *S. angustifolium* but is much larger. The widely spaced stem

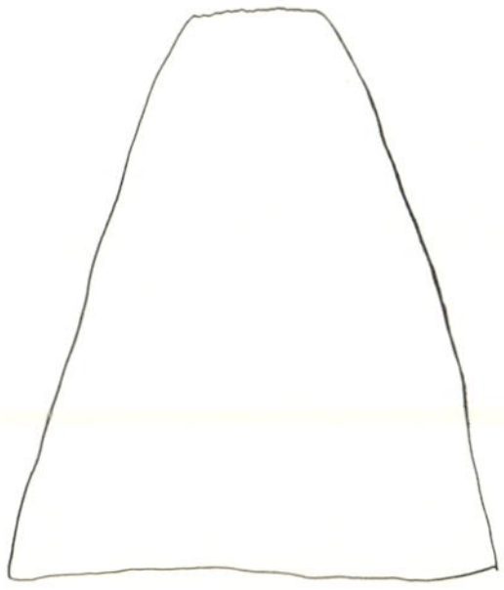

Figure 29. *Sphagnum recurvum* stem leaf.

leaves are usually longer than 1 mm and often have a ragged apex, which distinguishes it from *S. fallax*. The branch leaves are normally arranged in five rows on the branches, which is useful in distinguishing it from *S. flexuosum*. It is often found growing with *S. fallax*, *S. flexuosum*, *S. capillifolium*, and *S. papillosum*.

S. flexuosum Dozy and Molk.

(figure 30 and plates 23, 24, and 30)

Plants small to medium in size, soft looking, forming loose carpets and low hummocks near water level in open fens, on the edge of bog mats, and in damp woodlands. Plants are pale yellow-green to yellow-brown. Capitulum is small, the branches loose and twisted making it look like a loose ball of yarn. Stem is pale green to yellow. Branches arranged in fascicles of 2 + 2 (2 + 3, 2 + 1); spreading branches loose and tangled, branch leaves loose and spreading, pendent branches slender and tightly appressed to stem. Branch leaves are ovate-lanceolate (up to 1.5 mm long), undulate, and recurved when dry. Stem leaves are triangular-lingulate (0.8–1.2 mm long), with apex truncated or rounded and very ragged or torn looking.

This species often grows mixed in with *S. angustifolium*, *S. majus*, *S. subsecundum*, and *S. teres*. It is similar to *S. fallax*, but its stem leaves have a ragged apex. The branch leaves are not arranged in five rows, which distinguishes the species from *S. recurvum*.

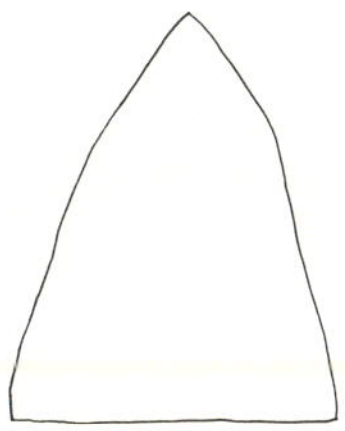

Figure 30. *Sphagnum flexuosum* stem leaf.

S. pulchrum (Braithw.) Warnst.

(figure 31 and plate 25)

Plants moderate in size, robust, densely branched, shiny when dry; forms dense, floating carpets in fens and beside pools near the margins of bogs. It is often a pioneer species on the margins of ponds. Plants are brownish or golden brown in the upper portions and yellow to dirty green in the lower portions. Capitulum is composed of short, thick, swollen branches. Stem is pale yellow, yellow-brown, to brown. Branches arranged in fascicles of 2 + 2; spreading branches thick and swollen, longer than pendent branches; pendent branches swollen and short (up to 1.3 cm long), loosely covering stem. Branch leaves are large (1.2–1.8 mm long), ovate to broadly ovate-lanceolate, concave, margins involute almost to middle from apex, slightly secund. Stem leaves broadly triangular (0.9–1.2 mm long), margins inrolled at apex to form abrupt, distinct point or cusp.

Figure 31. *Sphagnum pulchrum* stem leaf.

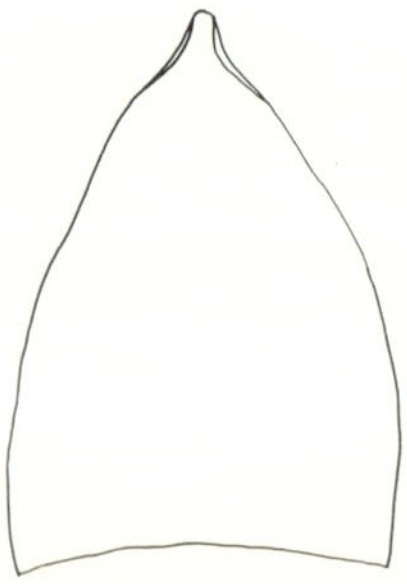

This plant often grows in association with *S. majus*, *S. fallax*, and *S. flexuosum*. Distinguishing characteristics are its dark brownish color, swollen, densely packed branches, and stem leaves with an abrupt point.

Sphagnum lindbergii, found in only a few places in the United States, can be confused with *S. pulchrum*, however *S. lindbergii* has five-ranked branch leaves, lacerated stem leaves much like *S. fimbriatum*, and a very dark brown or black stem.

S. fallax (Klinggr.) Klinggr.

(figure 32 and plate 26)

Plants robust and thickly branched, forming loose to dense carpets in relatively minerotrophic, open or shaded habitats such as hollows, wet depressions, or near the edges of pools. Plants are green, pale yellow-green, yellowish, often tinged with brown. Capitulum is stellate, bases of branches tinged with light red. Stem is pale green to yellowish. Branches arranged in fascicles of 2 + 2 (2 + 3); spreading branches slender and tapering (1.0–2.0 cm long), leaves arranged in five rows on branches; pendent branches appressed to stem and not much different from spreading branches. Branch leaves are ovate-lanceolate (1.0–2.0 mm long), undulate, and recurved at tips when dry. Stem leaves are triangular to triangular-ovate (0.7–1.1 mm long), with apex apiculated.

A dominant species of wet depressions in open bogs and fens, this plant often forms dense carpets. Although usually green, the more exposed plants may be yellow-brown. Submerged forms may resemble *S. cuspidatum*, but the short, straight branch leaves, undulate and recurved when dry, and the short triangular stem leaves with apiculated tips separate it easily from that species.

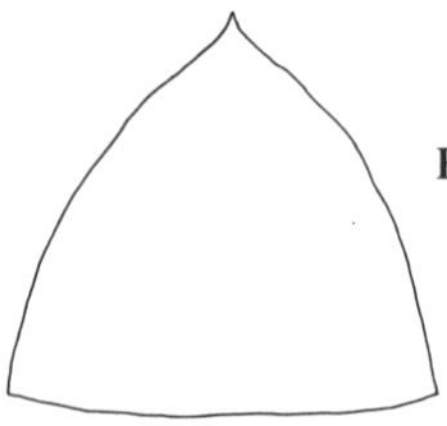

Figure 32. *Sphagnum fallax* stem leaf.

Macroscopic Key to All Species of *Sphagnum*

The following key was designed for identification of the majority of the species found in northeastern North America. Anisophyllous plants, plants with stem leaves that are maximally differentiated from branch leaves, were used to construct the key. For this reason, hemiisophyllous plants, those with stem leaves only partially differentiated from branch leaves, may not key out. Also, since the key was written using specimens from northeastern North America, specimens from other parts of North America may not key out. Some species, especially those in sections *Sphagnum* and *Subsecunda*, are difficult to separate even using microscopic characteristics. In spite of these limitations, people who have tested this key in different parts of North America and Europe have found it useful.

Key to Sections

1. Branch leaves cucullate
   2. Stem leaves small and triangular, less than 1 mm long; branch leaves somewhat squarrose—*Rigida*
   2. Stem leaves large, greater than 1 mm long, lingulate or fan-shaped; branch leaves tightly or loosely imbricate—*Sphagnum*
1. Branch leaf margins inrolled or flat, but not cucullate
   3. Six or more branches per fascicle—*Polyclada* (monotypic: *S. wulfianum*)
   3. Five or fewer branches per fascicle

4. Branch leaves squarrose
    5. Stem leaves small and triangular, less than 1 mm long—*Rigida*
    5. Stem leaves large and lingulate, 1–1.5 mm long—*Squarrosa*
4. Branch leaves not squarrose
    6. Extremely long, nonundulate, tufted branch leaves; plants dark brown, greenish, or silvery; primarily coastal plants—*Isocladus* (monotypic: *S. macrophyllum*)
    6. Coastal or continental plants with undulate or nonundulate branch leaves; plants various colors
        7. Hanging and spreading branches very similar, flexuose, about the same length; or plants with few or no branches—*Subsecunda*
        7. Hanging branches usually longer and more slender than spreading branches; the distinction between hanging and spreading branches is pronounced
            8. Stem leaves much smaller than branch leaves; plants various shades of green, yellow, or brown, never red, but sometimes branches and stems pinkish; on plants with five-part capitula, paired branches of fascicles present between rays of branches; plants of wet, mineral-rich depressions, submerged or near the water level—*Cuspidata*
            8. Stem leaves nearly the same size as branch leaves or larger; plants var-

ious shades of green, brown, or red; on plants with five-part capitula, branches of fascicles single or absent between rays of branches; plants of drier, mineral-poor, acid habitats, not normally submerged, usually growing above water level—*Acutifolia*

Section *Acutifolia*

1. Many fascicles with three spreading and two hanging branches—*S. quinquefarium*
1. Two spreading branches and one or two hanging branches per fascicle
    2. Stem leaves fan-shaped or lingulate and fringed at least at apex
        3. Stem leaves fan-shaped and fringed from apex down sides nearly to base; terminal bud large; often fruiting—*S. fimbriatum*
        3. Stem leaves lingulate and fringed only at apex or with distinct notch; terminal bud lacking or poorly developed; seldom fruiting
            4. Stem leaves fringed across apex; plants stiff and wiry with long, slender spreading branches perpendicular to stem; plants green, never red—*S. girgensohnii*
            4. Stem leaves with distinct notch at apex; plants not especially wiry or stiff; plants green, but usually splotched with red—*S. russowii*
    2. Stem leaves variously shaped, margins entire, never fimbriate, with rounded or pointed apex
        5. Capitulum flat, five-parted; branch leaves in five rows

- 6. Found in *Thuja*, *Abies*, and/or *Picea* swamps and rich minerotrophic fens; slender spreading branches with leaves in five distinct rows, leaves straight, not curved; stem leaves short with a length to width ratio of 1.5:1 or less, lingulate or triangular with apex usually rounded, sometimes pointed; plants green, but usually with a purplish red tinge, rarely red throughout—*S. warnstorfii*
- 6. Plants of open bogs, commonly forms hummocks; branch leaves usually in five rows, leaves usually curved to one side; stem leaves long lingulate, with a length to width ratio of 1.5:1 or longer, with rounded or pointed apex; plants usually red throughout or splotched red-green—*S. capillifolium* var. *tenellum*

5. Capitulum flat, rounded, or hemispherical, not five-parted; branch leaves not in five rows
   - 7. Plants greenish brown, yellow-brown, brown, or deep red-brown
     - 8. Stem leaves lingulate with broad, rounded apex; forms hummocks in dry, open, acid bogs—*S. fuscum*
     - 8. Stem leaves lingulate-triangular, apex not rounded; forms mats or small hummocks in wet minerotrophic habitats
       - 9. Plants large and stiff to touch; stem leaf length-to-width ratio is 1.5:1 or greater; stem leaf apex apiculate or

right-angled; branches mostly 1-mm wide or greater; usually in coastal habitats—*S. flavicomans*

9. Plants moderate-sized and soft to touch; stem leaf length-to-width ratio less than 1.5:1; stem leaves with obtusely angled apex; forms hummocks in shrubby or wooded fens—*S. subfulvum*

7. Plants not greenish brown, brown, or red-brown

10. Stem leaves short, length-to-width ratio less than 1.5:1; triangular; forms hummocks in damp, coniferous forests and hardwood swamps—*S. subtile*

10. Stem leaves long, length-to-width ratio greater than 1.5:1, lingulate, ovate, not triangular; plants of open bog habitats

11. Stem leaves broadest at base

12. Stem leaf margins near apex strongly inrolled, apex is toothed, stem leaves greater than 1.5 mm long; branches with swollen appearance; plants of coastal areas on damp sand or on pond margins—*S. tenerum*

12. Stem leaf margins not strongly inrolled, apex not strongly toothed, stem leaves less than 1.5 mm; branches do not appear swollen; forms hum-

mocks in open and treed acid bogs—*S. capillifolium* var. *capillifolium*

11. Stem leaves broadest above middle
    13. Stem leaf with truncate, dentate apex; branch leaves not strongly inrolled at apex, six to ten teeth at apex; plants soft and lax; capitulum has large, blunt, flat branches; plants usually pale green or purple-red—*S. angermanicum*
    13. Stem leaf with broad, toothed apex; branch leaves with inrolled margins, apex with up to eight large teeth; plants stiff when dry; capitulum large and compact with round, pointed branches; plants pale white, yellow, or purple, rarely deep colored—*S. molle*

Section *Cuspidata*

1. Stem leaves large, 1–1.5 mm long, apex irregularly cut or torn
   2. Stem leaves fan-shaped or squarish, broad apex cut or torn with coarse teeth; branch leaves in five rows; stem dark brown—*S. lindbergii*
   2. Stem leaves broad triangular, not fan-shaped; stem leaves torn down the middle; branch leaves usually not in five rows; plants pale

green or yellow, rarely dark brown—*S. riparium*

1. Stem leaves large or small, apex not irregularly cut or torn
   3. Stem and branch leaves almost the same size and shape; stem leaves lingulate; branch leaves strongly concave, not inrolled at apex, not undulate, not recurved—*S. tenellum*
   3. Stem and branch leaves distinctly different in size and shape
      4. Stem leaf apex truncate or broadly rounded, 0.8 mm or longer
         5. Branch leaves in five rows; hanging and spreading branches and leaves similar, hanging branches loosely cover the stem, branch leaves long and narrow, recurved at tips when dry; capitulum five-parted—*S. recurvum*
         5. Branch leaves not in five rows; hanging and spreading branches different, with hanging branches tightly covering the stem; branch leaves ovate to broadly ovate-lanceolate; the capitulum resembles a small, loose ball of yarn—*S. flexuosum*
      4. Stem leaf apex acute, apiculate, or rounded, but not truncate, 0.5 mm or longer
         6. Apex of stem leaf apiculate
            7. Branches short and thick with broad, apiculate leaves, branch leaves in five rows; spreading and hanging branch leaves different; plants green, yellow, and brown—*S. pulchrum*

7. Spreading and hanging branches slender, branch leaves not in five rows; spreading and hanging branch leaves similar; plants green, yellow, seldom brown—*S. fallax*

6. Stem leaves with acute or rounded apex, not apiculate

8. Stem leaves small, less than 0.8 mm long; plants usually above water level in acid bogs and poor fens—*S. angustifolium*

8. Stem leaves longer than 0.8 mm; plants primarily aquatic or found in very wet, mineral-rich habitats

9. Stem leaves with rounded apex

10. Plants brownish green, almost black; branch leaves usually in five rows and are strongly curved and turned to one side at branch tips; weakly minerotrophic, aquatic or near water level in wet depressions—*S. majus*

10. Plants orange-brown, yellow-brown; branch leaves not in five rows, leaves curved at branch tips, but not all turned in the same direction; minerotrophic and aquatic—*S. jensenii*

9. Stem leaves with acute apex

11. Branch leaves long, slender, and inrolled only near apex; branch leaves not in five rows; teeth on branch leaf margin near apex; plants are plumose;

grows primarily in coastal regions—*S. trinitense*

11. Branch leaves long, slender, and inrolled at least half the length of the leaf; branch leaves usually in five rows; branch leaf margins without teeth; plants not plumose; near coast and/or inland
    12. Branch leaves inrolled from middle to apex, narrow, often 1 mm wide; sides of triangular stem leaves longer than base; plants slender, weak, and floppy; capitulum not noticeable; plants green or yellow, bases of branches in capitulum may be red-tinged—*S. cuspidatum*
    12. Branch leaves inrolled from near base to apex, usually less than 0.8-mm wide; base and sides of triangular stem leaves are almost equal in length; plants large; capitulum noticeable; plants green, yellow, or brown—*S. torreyanum*

Section *Isocladus*

Extremely long, nonundulate, tufted branch leaves; plants dark brown, greenish, or silvery; primarily coastal plants; could be confused with dark aquatic forms of *S. torreyanum*, but its branch leaves not as tufted—*S. macrophyllum*

Section *Polyclada*

Six or more branches per fascicle; branch leaves in five rows; hemispherical capitulum; thick stems sometimes dark brown to black; large plants forming loose carpets on damp humus in mineral-poor, coniferous swamps and fens—*S. wulfianum*

Section *Rigida*

1. Stem leaves of two types: either triangular with short slit down middle at apex or oval-shaped with rounded apex; branch leaves with rounded apex, margins inrolled; stem brown, plants pale brownish green, sometimes tinged with violet—*S. compactum*
1. Stem leaves triangular with broadly rounded apex, almost truncate; branch leaves elongated, leaves abruptly inrolled above middle to form long, tubular apex; stem pale green, plants whitish green—*S. strictum*

Section *Sphagnum*

1. Stem leaves long rectangular; plants pink and/or purple-red in open areas and green in shaded areas—*S. magellanicum*
1. Stem leaves not long rectangular; plants green, yellow, brown, but not red
    2. Stem leaves fan-shaped, greatly expanded below apex; two to three spreading branches and three hanging branches; plants green, gold-brown, purple-brown; forms dense carpets in wet forests—*S. centrale*
    2. Stem leaves oval-shaped, leaves widest at middle
        3. Branches, especially in the capitulum, slen-

derly club-shaped; plants green, brown, or gold-brown; coastal habitats

4. Growing submerged or near water level in shallow, sandy-bottomed ponds along shorelines in shade or sun; plants lax and weak stemmed, forms loose carpets in shade or sun—*S. portoricense*
4. Forms large hummocks in full sunlight in ombrotrophic to weakly minerotrophic peatlands; plants compact in large hummocks—*S. austinii*

3. Branches not club-shaped; plants usually not growing submerged; not strictly coastal
   5. Capitulum small and distinctly flattened with small, pointed branches; plants green, reddish brown in capitulum; newest branches of capitulum noticeably shorter than branches below—*S. henryense*
   5. Capitulum large and rounded; plants green, brown, purple-brown; newest branches not noticeably shorter than older branches
      6. Stem leaves greater than 1.5 mm long and length is almost twice the width; plants green to gold-brown; grows on margins of mineral-rich habitats, rarely dominant—*S. palustre*
      6. Stem leaves less than 1.5 mm long, only slightly longer than broad
         7. Stem leaf margins entire, sides nearly parallel; branches spreading, short, thick, and blunt; plants gold-brown to purple-brown,

rarely green; found on wet carpets in open areas, sometimes on rocks in mountains, but rarely found in forests—*S. papillosum*

7. Stem leaf margins slightly ragged above middle, apex very broadly rounded, stem leaf almost fan-shaped, widest in middle; branch leaves frequently squarrose; plants slender overall; plants green, tinged brown or purple-brown; when growing intermixed with other species of this section, this species is paler—*S. imbricatum* (*S. affine*, a very similar species, is virtually indistinguishable macroscopically)

Section *Squarrosa*

1. Branch leaves strongly squarrose throughout plants; stem leaves long triangular with truncate apex; pale green, rarely dark brown—*S. squarrosum*
1. Branch leaves imbricate in upper portion of plant and slightly squarrose in lower portion; stem leaves tongue-shaped with rounded or flat apex, apex may have three to four low, broad teeth; plants brown in open areas, pale green in shaded areas—*S. teres*

Section *Subsecunda*

1. Plants slender, unbranched or with one to two branches per fascicle, capitulum absent—*S. pylaesii*

1. Plants large, usually branched, capitulum present
   2. Stem leaves 0.5–1.0 mm long; branch leaves slightly curved to one side
      3. Stem leaves only slightly longer than broad, with broad apex, margin flat; branch leaves not much longer than stem leaves, margins inrolled nearly entire length, leaves slightly curved to one side; small, slender, wiry plants; capitulum usually without a terminal bud; stem dark brown—*S. subsecundum*
      3. Stem leaves nearly twice as long as broad, slightly rounded at apex, nearly acute, margins flat or inrolled; branch leaves nearly twice the length of the stem leaves, inrolled only near the apex, slightly curved to one side; capitulum large, flat, terminal bud present; stem pale green to light brown—*S. contortum*
   2. Stem leaves longer than 1.0 mm; branch leaves rarely curved to one side
      4. Terminal bud large; one to three branches per fascicle, two spreading and one hanging, little or no difference between hanging and spreading branches; branches do not conceal stem; stem and branch leaves of similar size, shape, and structure—*S. platyphyllum*
      4. Terminal bud indistinct; four to five branches per fascicle, two to three spreading and two hanging, branches noticeably different; hanging branches conceal stem; stem and branch leaves noticeably different in size, shape, and structure—*S. lescurii*

Glossary

Acute. Sharply pointed, less than 90 degree angle.

Anisophyllous. Plants with stem leaves maximally differentiated from branch leaves. Often an indication of optimal habitat conditions.

Antheridia (sing. *antheridium*). Sperm-bearing structures found at the base of leaves in the upper portion of a gametophyte. These branches are usually club-shaped and frequently tinged with some color other than green.

Apex. The tip.

Apiculate. Ending in an abrupt, short, sharp point.

Archegonia (sing. *archegonium*). Egg-bearing structures found at base of fascicles in upper portion of plant.

Bog. Wetlands that are very poor in nutrients, most of the minerals being obtained from precipitation. The dominant vegetation is *Sphagnum*.

Branch leaves. Leaves on branches, usually different in shape from stem leaves.

Capitulum (pl. *capitula*). The top of a *Sphagnum* plant composed of small branches.

Capsule. The spore case of a sporophyte.

Carr. A minerotrophic peatland containing small trees and shrubs such as *Alnus* and *Salix* species.

Circumpolar. Widespread around hemisphere.

Crispate, crisped. Wavy, bent, curled, contorted, or twisted.

Cucullate. Hood-shaped—margins at apex of leaf curled in to form a hood.

Dentate. Having large, triangular, outward pointing teeth on margins and tips of leaves.

Dioicous. Having antheridia and archegonia on separate plants.

Diploid. Having two sets of chromosomes. Characteristic of the sporophyte phase of a plant life cycle.

Entire. Smooth, without teeth. Usually refers to margin of leaf.

Falcate. Sickle-shaped.

Falcate secund. Sickle-shaped leaves all bent in the same direction on a branch.

Fascicles. Group or bunch of branches connected at the base. Most *Sphagnum* species have two kinds of branches in each fascicle—pendent and spreading branches.

Fen. Peat-forming wetlands that receive some nutrients from the surrounding ground water.

Fimbriate. Minutely fringed.

Five-parted. Refers to capitulum in which five rays of branches radiate out from center like spokes of a wheel; also called stellate.

Five-ranked. Leaves arranged in five distinct rows on a branch.

Flexuose. Wavy, undulate, or slightly bent, curved, or twisted.

Fruiting. Refers to gametophyte with attached sporophytes.

Gametophyte. The dominant phase of the life cycle of *Sphagnum*—the leafy plant.

Hanging branch. A vertical hanging branch of a fascicle, usually more slender than the horizontal spreading branches of a fascicle and usually appressed to the stem.

Haploid. Having one set of chromosomes. Characteristic of the gametophyte phase of a plant life cycle.

Hemiisophyllous. Plant with stem leaves only partially differentiated from branch leaves. Usually indicates some periodic fluctuation of water level around plant.

Hollows. Small, wet depressions at the base of hummocks.

Hummock. A low mound composed of *Sphagnum*.

Hydrarch Succession. Filling in of a body of water by peat, with eventual succession to a forest.

Hydrophytic. Typically found in wet habitats.

Hygrophytic. Typically found in damp habitats.

Imbricate. Leaves appressed and overlapping like shingles on a roof.

Incurved. Margins of leaf curved upward and inward.

Involute. Margins of leaf rolled inward.

Ion. A positively or negatively charged atom.

Isophyllous. Plant with stem leaves identical to branch leaves. Usually indicates frequent fluctuation in water level around plant.

Lacerate. Jagged, cut, or torn in appearance.

Lagg. Mineral-rich drainage area surrounding a peatland where minerals are eroded from the adjacent upland, which usually has a flora more characteristic of a richer habitat such as a marsh.

Lanceolate. Lance-shaped, tapering at both ends.

Lingulate. Tongue-shaped, long with parallel sides and broad apex.

Lingulate-triangular. Two sides much longer than base.

Margins. Edge of leaf.

Marsh. Grass and/or sedge wetland with standing water throughout most of the year.

Minerotrophic. Rich in minerals.

Monoicous. Plants with antheridia and archegonia on the same plant.

Ombrotrophic. Receiving nutrients solely from precipitation. A true bog.

Open bog. Very few trees present.

Ovate. Egg-shaped, widest below the middle.

Paludification. Flooding of an upland; as it applies to peatlands, the flooding of an upland by rising water table due to peat accumulation.

Peat. Soil with an organic content of at least 20 percent and an ash content of less than 50 percent that develops under water-logged, anaerobic conditions.

Peatland. A wetland that accumulates peat.

Pendent branch. Same as hanging branch.

Plant. For *Sphagnum*, the gametophyte or dominant phase of the life cycle.

Perichaetial leaves. Leaves that surround the archegonium. These are usually larger than the other branch leaves.

Poor fen. Wetlands receiving some minerals from ground water and dominated by *Sphagnum*.

Protonema (pl. *protonemata*). The early filamentous

phase of the gametophyte produced by germination of a spore. Protonemata may also develop from gametophyte fragments of a mature plant.

Rays. Groups of branches radiating from center of capitulum. When present, there are usually five rays in a capitulum.

Recurved. Bent backwards in an even curve. Recurving is not noticeable until the plant is dry. See "squarrose."

Rich fen. Wetland receiving minerals from ground water, dominated by sedges with few peat mosses.

Section. A subdivision of closely related species within a genus.

Secund. Turned to one side.

Serrate. Saw-toothed, with small, sharp, forward-pointing teeth.

Sporophyte. The spore-producing phase of the life cycle, arising from a fertilized egg and raised above the plant on a green stalk, the pseudopodium.

Spatulate. Fan-shaped. Tapering from a broad, rounded apex to a narrower base.

Spore. Haploid cell produced by sporophyte, capable of developing into a new gametophyte.

Spreading branch. The horizontally oriented branches of a fascicle, usually larger than the pendent branches.

Squarrose. Upper portion of leaf bent back towards the base of the leaf in a sharp, right-angled bend. Squarrosity is present when the plant is wet or dry. See "recurved."

Stellate. Star-shaped. Refers to capitulum in which five rays of branches radiate out from center—like spokes of a wheel.

Stem leaves. Leaves on main axis of plant. These are usually different in shape from branch leaves.

Swamp. A woodland with standing water for most of the summer.

Terminal bud. Growing tip in the middle of the capitulum. Also known as apical bud. In some species, it is very noticeable, such as *S. fimbriatum* and *S. riparium*.

Terete. Refers to branches, cylindrical in shape, with the leaves pressed against one another.

Triangular-lingulate. Two sides longer than base.

Truncate. Ending abruptly or appearing to be cut off squarely at the apex.

Tufted. Refers to an erect group of structures. A group of more or less erect shoots, branches, or leaves.

Undulate. Wavy.

Further Reading

Andrus, R. E. 1980. Sphagnaceae *(Peat Moss Family) of New York State*. Bulletin no. 442. New York State Museum, Albany, New York. (An excellent, inexpensive book that includes nearly all the species of North America and an excellent discussion of the ecology of peat mosses.)

Crum, H. 1984. *Sphagnopsida*. Sphagnaceae. North American flora, series II, part 11. New York Botanical Garden, Bronx, New York. (A moderately expensive book with excellent illustrations of all the species in North America.)

Crum, H. 1988. *A Focus on Peatlands and Peat Mosses*. The University of Michigan Press, Ann Arbor, Michigan. (A moderately expensive book with an overview of peatland ecology. Keys and illustrations to common species of North America are included, with illustrations and keys from the previous publication. Crum has also included many photomicrographs of morphological features for each species. The keys are designed for use with a compound microscope, however.)

Crum, H., and L. E. Anderson. 1981. *Mosses of Eastern North America*. Columbia University Press, New York. (This two-volume work is rather expensive, but it includes all other groups of mosses. The brief descriptions and few drawings are of limited value unless one has a microscope.)

Ireland, R. R. 1982. *Moss Flora of the Maritime Provinces*. National Museums of Canada, publications in botany no. 13, Ottawa, Canada. (This relatively inexpensive book has full-page illustrations, includes the other groups of mosses, and discusses collecting and preserving specimens for permanent collections.)

Index